The Publication of

A County Built on Iron

St. Louis County, Minnesota

1856-2006

was made possible through the generous sponsorship of

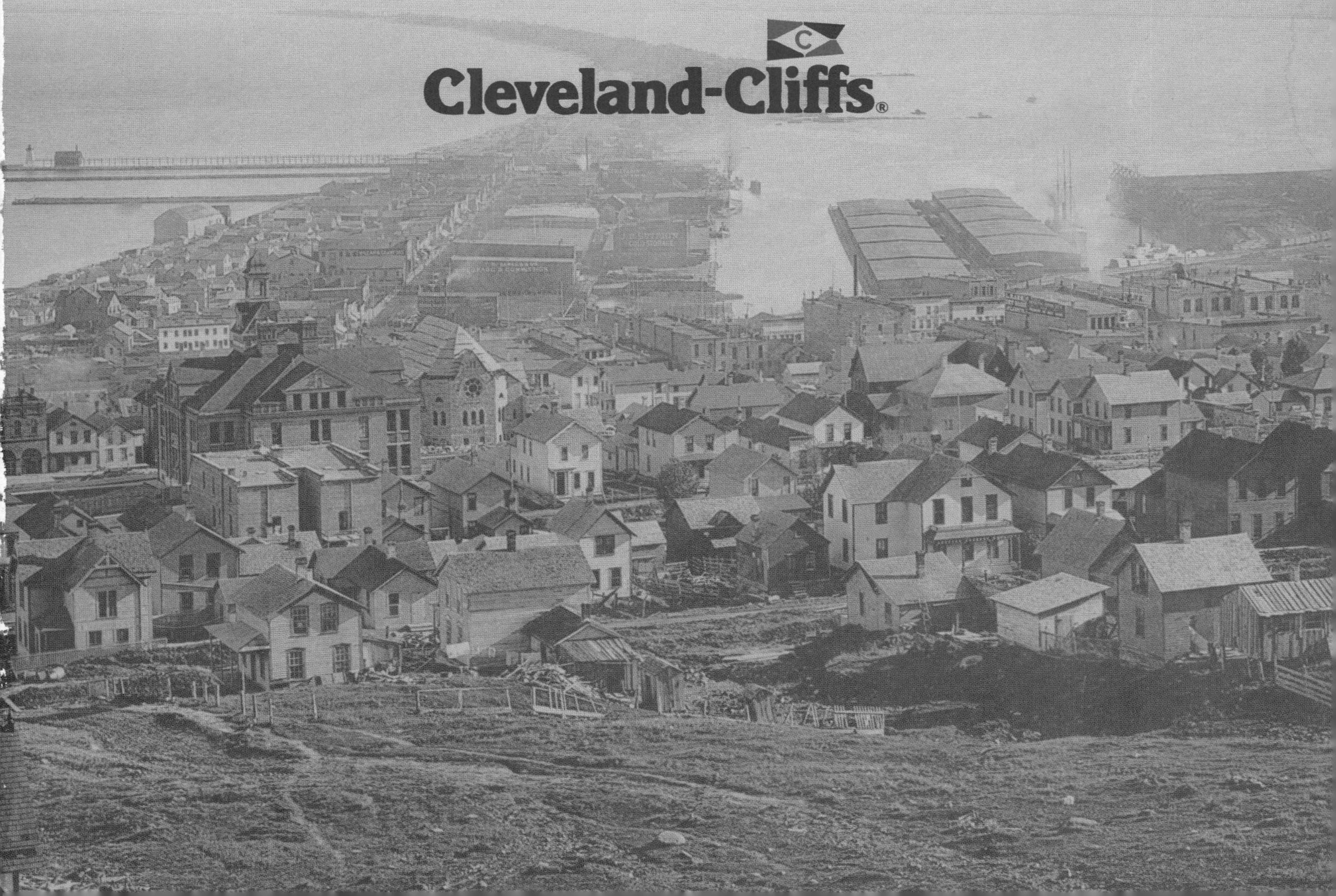

THE
DONNING COMPANY
PUBLISHERS

A County Built on Iron

St. Louis County, Minnesota

1856–2006

by Bill Beck

The Donning Company Publishers
184 Business Park Drive, Suite 206
Virginia Beach, VA 23462

Steve Mull, General Manager
Barbara B. Buchanan, Office Manager
Richard A. Horwege, Senior Editor
Lynn Parrott, Graphic Designer
Amy Thomann, Imaging Artist
Lori Kennedy, Project Research Coordinator
Scott Rule, Director of Marketing
Stephanie Linneman, Marketing Coordinator

Steve Mull, Project Director

Library of Congress Cataloging-in-Publication Data

Beck, William O.
A county built on iron : St. Louis County, Minnesota, 1856–2006 / by William O. Beck.
p. cm.
Includes index.
ISBN-13: 978-1-57864-342-4 (hard cover : alk. paper)
ISBN-10: 1-57864-342-2 (hard cover : alk. paper)
1. Saint Louis County (Minn.)—History. 2. Saint Louis County (Minn.)—Biography. 3. Saint Louis County (Minn.)—Economic conditions. 4. Iron industry and trade—Minnesota—Saint Louis County—History. 5. Iron mines and mining—Minnesota—Saint Louis County—History. I. Title.
F612.S25B43 2006
977.6'77—dc22

2005038039

Printed in the USA at Walsworth Publishing Company

Contents

128

91

15

54

Foreword

The face of a region changes inevitably over time. We have often wondered whether the tourists who throng Canal Park in Duluth on summer afternoons to watch the thousand-footers glide beneath the Aerial Bridge realize that they are witnessing the contemporary chapter of a pageant that dates back a century-and-a-half.

Commercial shipping on Lake Superior got its start in 1855 when the U.S. government opened the Soo Locks to allow vessels to bypass the rapids of the St. Mary's River at Sault Ste. Marie. The very next year, the then Territory of Minnesota carved out a county from a huge block of land in northeastern Minnesota and gave it the name of a medieval French king. St. Louis County was little more than wilderness then, with only a handful of pioneers grubbing out a living at the mouth of the St. Louis River.

The county's early years were characterized by boom and bust, with the Panic of 1857 followed by the coming of the railroad in 1871—and the digging of that ship canal that tourists still find so fascinating today—which was in turn followed by the Panic of 1873. Some of those early settlers could have been forgiven if they thought St. Louis County might never amount to much.

But the 1880s and 1890s revealed that the hills north and west of Lake Superior concealed one of the globe's richest deposits of iron ore. In the space of little more than twenty years, St. Louis County became one of the fastest-growing counties in the Midwest as thousands of immigrants flooded into Duluth and the infant cities of the Mesabi Range, drawn by abundant jobs in the iron mining, timber harvesting, and commercial shipping industries. The development of the region's natural resources economy at the same time that America became a beacon for democracy gave St. Louis County its rich heritage.

The face of the land we call home has changed dramatically in these past 150 years. Open pits that are visible from space are visible evidence of the county's mining history, but in another 150 years, most of those pits—reforested by the mining companies and Iron Range Resources—will appear little different from the natural lakes that dot the county. And a century after the last White Pine was cut in northeastern Minnesota, an unbroken stretch of boreal forest carpets the county from its southern border north to Canada.

St. Louis County, 150 years after its formation, is a vibrant community that still relies upon its natural resources for economic sustenance. But the county also boasts a thriving tourism industry, outstanding secondary education opportunities for residents and outsiders alike, and a quality health care industry in Duluth and the Iron Range communities of Virginia and Hibbing that attract patients from across northern Minnesota and the western Lake Superior basin.

The story of the county's development these past 150 years is, we hope, a fascinating read. It is an account of challenge and opportunities, boom and bust, and victory and defeat, a chronicle of an intrepid citizenry who carved a county out of the wilderness and prepared this place we call St. Louis County for a bright future in the twenty-first century.

Robert Mars, President
JoAnne Coombe, Executive Director

Acknowledgments

A project as intrinsically complex as writing the history of a county that is celebrating its 150th birthday involves far more than the talents of just one individual writer. *A County Built on Iron* was a massive undertaking, but an outgrowth of several books I have written on the history of St. Louis County and northeastern Minnesota. Research is a cumulative effort, and materials I consulted more than twenty years ago for *Northern Lights*, my history of Minnesota Power, were put to good use for this project. Similarly, the massive amount of work that went into *Pride of the Inland Sea*, my 2004 history of the Port of Duluth-Superior, paid dividends during the research phase of *A County Built on Iron.*

Twenty years of writing books has taught me one enduring lesson. There are dozens of people on each and every project whose support and friendship make a book project possible. Writing books is like going into battle. Every front line military unit has a supply and logistics "tail" that makes it possible to sustain battle.

JoAnne Coombe and Bob Mars wanted this story told. I first met Bob a quarter-century ago when I was a reporter for the *Duluth News-Tribune and Herald.* Bob was a longtime board member at Minnesota Power when I worked in the utility's corporate public affairs department during the 1980s. We renewed our acquaintance last year with the St. Louis County history project and a centennial history of Minnesota Power, which will be published in the spring of 2006. Bob was one of the readers of the manuscript whose invaluable suggestions are incorporated in these pages.

I've known and admired the work of JoAnne Coombe, the Society's executive director, since I myself was a member of the Society's board some twenty years ago. JoAnne runs one of the best historical societies in the state of Minnesota and is always a joy to work with.

Pat Maus never ceases to amaze me. The archivist for the Northeast Minnesota Historical Center at the University of Minnesota Duluth, Pat has been collaborating with me on books since *Northern Lights.* Pat's knowledge of county history is truly encyclopaedic. And all of the beautiful illustrations you see in *A County Built on Iron* were selections that Pat made from the Center's comprehensive collection of historical photographs.

Larry Sommer, the recently retired executive director of the Nebraska Historical Society and JoAnne's predecessor as executive director of the St. Louis County Historical Society, shared his insights into county history during a long luncheon visit in Indianapolis this past summer.

Dick Hudelson at the University of Wisconsin-Superior read the manuscript from first page to last. His background and expertise in the county's rich labor and cultural history was valuable in lending balance to the story.

Davis Helberg, my collaborator on *Pride of the Inland Sea*, is a friend of more than a quarter-century and a priceless repository of the county's maritime history. Larry and Jerilyn Fortner made the church in West Duluth and the cabin on Scrapper Lake available for a writer in need of safe refuge. Wayne Brandt has always shared his knowledge of the region's timber harvesting culture, and Jack LaVoy of Iron Range Resources imparted his knowledge of the area's more recent economic development history during a leisurely lunch at the Pickwick.

Publisher Steve Mull at Donning served as cheerleader and taskmaster, and Editor Richard Horwege kept the project on task and on schedule.

Elizabeth, my wife and copy editor, remains my inspiration four decades after our first dance at the Beanie Bounce. None of this would be possible without her.

Finally, the inevitable omissions, misinterpretations, and errors contained within are on my shoulders, and my shoulders alone.

Bill Beck
Indianapolis, Indiana
November 13, 2005

Creating a County from the Wilderness

chapter one

A century-and-a-half ago, there was only wilderness. The land that we know as St. Louis County was uncrossed by roads or populated by cities. The few inhabitants lived in primitive cabins, mostly along the lower St. Louis River where it emptied into Lake Superior—which technically made most residents of the county in 1856 citizens of Wisconsin.

Although few roads led into the interior of the sprawling county, the region did boast a transportation system that had been in place since Native Americans first penetrated the headlands west of Lake Superior eons ago. The rivers and streams that bisected the county created a water highway in use for thousands of years. Native Americans paddled across North America in canoes made of birch bark. Journeys that might take overland travelers months were accomplished in days on the swift-flowing streams of the Upper Great Lakes.

The Great Lakes basin and Canada contain well over half the world's freshwater surface.[1] When the last glaciers melted and retreated north some ten thousand years ago, they left behind a landmass between the Great Lakes and the Rocky Mountains that was intersected by countless waterways. A paddler entering the estuary of the St. Louis River, through the judicious use of portages, could travel north to Hudson Bay via Lake Winnipeg and the Nelson River. A short portage from Lake Winnipeg to Lake Winnibegosis carried the traveler to the foothills of the Rocky Mountains via the Saskatchewan River.

A century-and-a-half ago, there was only wilderness. The land that we know as St. Louis County…

Other portages north and west of Lake Winnipeg led to the Athabasca River, Great Slave Lake, and the Arctic Ocean via the mighty Mackenzie River. The long Savanna Portage west of the mouth of the St. Louis connected the water traveler with the

The first residents of northern Minnesota flesh out a wooly mammoth ten thousand years ago. The retreat of the last glaciers five to ten thousand years before created a massive glacial Lake Agassiz across much of what is today Minnesota and Manitoba. When the glacial lake finally receded, it left behind the Red Lakes in Minnesota and Manitoba's Lake Winnipeg. Primitive hunters made their way into the land uncovered by the shrinking glacial lakes and left behind stone implements that have been unearthed at numerous places in St. Louis County, including Island Lake. The Story of Minnesota's Past, p. 57

Upper Mississippi River system and the Father of Waters that drained a continent south to the Gulf of Mexico.

"Long before the Vikings landed at l'Anse aux Meadows or Columbus entered the Caribbean," wrote Minnesota fur trade archaeologist Douglas Birk, "diverse Indian groups likely understood the small craft-use potential of virtually all interior waterways in North America."[2]

The small craft in use was made from the bark of *betula papyrifera,* the paper birch that covered much of North America from the Great Lakes east to the Atlantic Ocean. Lightweight and maneuverable, the birch bark canoe could be built entirely from materials gathered in the northern forest.

Bark, preferably about a quarter inch thick, was peeled from a small tree for the outside covering of a canoe. Eastern white cedar logs were split and used for the gunwales, ribs, and sheathing. Native Americans stripped spruce roots and used the sinew remaining to lash the canoe together and to secure the birch bark to the frame. The sap from spruce trees was heated into a gum and used to seal the seams in the canoe.[3]

The resulting watercraft revolutionized travel in North America as early as several thousand years ago. When French peasants in the Dark Ages were consigned to living their entire lives never traveling more than five miles from their home village, Native Americans routinely were traversing a continent, from the Atlantic to the Pacific, from the Arctic Ocean to the Gulf of Mexico.

It was a lesson in transportation that wasn't lost on the first French explorers of North America four centuries ago.

St. Louis County under the Fleur de Lis

Samuel Champlain claimed North America for the king of France in 1608, at about the time English colonists set foot

Top: Eastman Johnson's 1857 oil painting Canoe of Indians *depicting the larger style of canoe employed by the Anishinabe of the Lake Superior region on lakes and larger rivers.* St. Louis County Historical Society Museum

The coureurs du bois who ranged across a continent in search of furs typically traveled in Montreal Canoes. Here, Radisson and Groseilliers, imagined by a modern artist, scan the shoreline of a Minnesota lake. Rhoda Gilman, The Story of Minnesota's Past, p. 67

on the Atlantic shore of a new continent at Jamestown and Plymouth Rock.

Champlain had founded a city on the St. Lawrence River at Quebec City, and next sent explorers upriver to solidify French control of the fertile river valley. The French explorers were soon joined by *voyageurs,* fur traders who ranged far into the interior in search of beaver and other pelts, as well as Jesuit and Recollet missionaries.

Both classes of French Canadian society almost immediately interacted with the continent's Native American population, the *voyageurs* for commercial reasons and the missionaries to save souls. Others ventured west from the settlements of *Nouvelle France* looking for the fabled Northwest Passage to Asia.

From the Lachine Rapids above Montreal, generations of *voyageurs,* also known as *coureurs de bois,* or wood rangers, priests, and soldiers headed west into the vastness of the North American continent, seeking gold, furs, salvation, or a mixture of all three. Most of those who headed into the wilderness in the 1600s were second or third sons. The France of the Bourbon kings was firmly rooted in the medieval concept of primogeniture. Since the eldest son inherited everything, he typically had little reason to abandon the old country for a life among what the colonial French would call the "savages" of New France.

Pierre Espirit Radisson and Medard Chouart, Sieur de Groseilliers, were two younger sons who made their names in the new world. Known to several generations of Canadian schoolchildren as "Radishes and Gooseberry," the two

For most of the two centuries that the fur trade reigned supreme in the Upper Great Lakes, the voyageurs *of* Nouvelle France *treaded gingerly across the vast expanses of the Great Lakes. Standard procedure called for following the shorelines, and only if the day was clear and the winds light would the Montreal Canoes of the fur brigades even consider crossing the mouth of a bay on Lake Superior. Here, in Frances Anne Hopkins'* Canoes in a Fog, Lake Superior, *a line of Montreal Canoes hug the shore of the big lake.* MHS Press, Spring 1999 Catalogue

intrepid explorers penetrated the Lake Superior country in the late 1650s. Then known as *Lac Tracy* to the French and *Kitchi-Gammi* to the Hurons and Anishinabe who populated its shores, the greatest of the Great Lakes was the key to penetrating the trade routes into the interior of North America.[4]

During the one hundred years after Radisson and Groseilliers first spied what would become known as *Lac Superieur*—the Upper Lake—*voyageurs* blazed three passages north and west off the big lake's western shore. At what is today Thunder Bay, they followed the Kaministikwia River west to the Rainy River and Lake of the Woods. A more southerly route followed the Pigeon River west after an arduous uphill portage at what is today Grand Portage near the United States–Canada border. A third route followed the St. Louis River west and north to long portages that took the *voyageur* north to the Hudson Bay country or south to the Mississippi River.

Others quickly followed Radisson and Groseilliers, who would transfer their expertise and allegiance to the English crown after a dispute with the officials in Montreal over sixty thousand livres worth of confiscated furs. Father Claude Dablon established the first mission on the big lake in 1659, on the south shore at La Pointe, sheltered by Chequamegon Bay.

Next to visit the area was Daniel Greysolon, Sieur du Lhut. Born at Saint-Germain-Laval, France, in the late 1630s, Greysolon was a career soldier for his king before leaving to seek his fortune in the new world during the late 1670s. He was preceded to New France by his cousins, the Tontys, who also would mark a legendary trail across North America in the 1600s.[5] In 1678, like so many before him, du Lhut dipped his paddle in the St. Lawrence and headed west. He wintered somewhere on the big lake, and in the spring of 1679, landed at what is now Fond du Lac.

Sieur du Lhut was the first European-American to step ashore on what would become St. Louis County more than three hundred years later. He laid claim to much of what is now Minnesota and the Dakotas, and dedicated his discovery to

St. Louis, the French king who had died fighting the Muslims in the crusades of the thirteenth century. He also lent his name to the city that would spring forth along the mouth of the St. Louis River.

The peripatetic du Lhut wouldn't stay at the Head of the Lakes. He explored west and north and is credited with building the fortified post of Kaministiquia on Thunder Bay and exploring the North Shore country surrounding Lake Nipigon. He rescued Father Louis Hennepin from the Dakota and fought the Iroquois—the hereditary enemies of the French—under La Barre and Denonville in the late 1680s. In 1686, on duty against the Iroquois, du Lhut helped lay the foundation of the critical post of Detroit.[6]

Top: This map depicts the route that Daniel Greysolon took during his journey through the western Lake Superior region between 1678 and 1681. NEMHC Pamphlet 326, p. 10

The house in which Daniel Greysolon, Sieur du Lhut, was born in the small town of St.-Germain-Laval, France. NEMHC Pamphlet 326, p. 8

Daniel Greysolon died in 1710 and was laid to rest in the Church of the Recollects in Montreal. Although he died broke, Greysolon was one of those adventurers who held New France for king and country for more than thirty years.

The Bay

For much of that period, *Nouvelle France* was sandwiched between English-speaking empires to the north and to the south, a geopolitical rock and a hard place that bedeviled the French-speaking colony for a century. The Hudson's Bay Company had been chartered by the king of England in 1670 to control the trade routes north and west of the Great Lakes. Hudson's Bay Company traders established what they called factories and *de facto* military posts on Hudson Bay and James Bay in an attempt to keep the French bottled up in the St. Lawrence River Valley. Hudson's Bay Company was the rock. For French Canada, the hard place was the increasingly aggressive English colonial empire south of the St. Lawrence.

From 1670 until 1760, Great Britain and France—through their commercial, colonial, and Native American allies in North America—battled for control of the fur trade in the heart of North America. The five nations of the Iroquois engaged their blood feud with the French through their British allies.

The fight was over a forty-pound rodent with lustrous fur that lent itself particularly well to men's fashions in Western Europe during the seventeenth and eighteenth centuries. The beaver, known as *ahmik* in a Native American language of the western Great Lakes, produced a soft fur that was used to manufacture coats and hats. For nearly two hundred years, beaver hats in a variety of styles were the height of fashion among European males.[7]

Daniel Greysolon, Sieur du Lhut, seen here landing near the St. Louis River in 1679, was on a mission from the governor of New France to open trade with the Lakota. During his visit to the Head of the Lakes, Greysolon ventured as far inland as Lake Mille Lacs, where he reportedly met up with Father Louis Hennepin, who had been briefly imprisoned by the Lakota. Rhoda Gilman, The Story of Minnesota's Past, p. 68

The French Canadian *voyageurs* and Hudson's Bay factors traded Native Americans a variety of cheap metal and textile goods for processed beaver pelts. The Native Americans considered their end of the deal by far the best. "The beaver does everything well," a Native American marveled to a *coureur du bois.* "He makes us kettles, axes, swords, knives and gives us drink and food without the trouble of cultivating the ground."[8]

The battle between Great Britain and France for control of the North American continent was decided less than a half-century after Daniel Greysolon's death when Quebec City fell in 1759 to the forces of James Wolfe on the Plains of Abraham outside the walls of the fortress city. In the bigger picture, the fall of New France was a somewhat minor episode in what the Europeans would come to call the Seven Years' War, a global conflict between Great Britain and France that would culminate in the Napoleonic wars a half-century in the future.[9]

The two major routes into the interior of North America from western Lake Superior were the Kaministikwia River west from the present site of Thunder Bay, Ontario, and the Pigeon River across the Height of Land from present-day Grand Portage, Minnesota. A third route, up the St. Louis River from Fond du Lac, took the traveler north to Rainy Lake and Lake of the Woods or south to the Mississippi River, both by arduous portages. St. Louis County has the unique distinction of containing three watersheds. A drop of water falling on the northern section of the county is likely to flow in one of three distinct directions: east to the Atlantic Ocean via the Great Lakes; south to the Gulf of Mexico via the Mississippi River; and north to Hudson Bay via Rainy Lake. Robert C. Wheeler, Toast to the Fur Trade, p. 28

Fond du Lac, shown here in an 1827 view, was so admirably situated astride the fur trading routes of the interior of North America that it served as a trading post under the three regimes that governed the continent between 1700 and 1840: New France, Great Britain, and the United States. Thomas L. McKenney, Tour to the Lakes, 1827

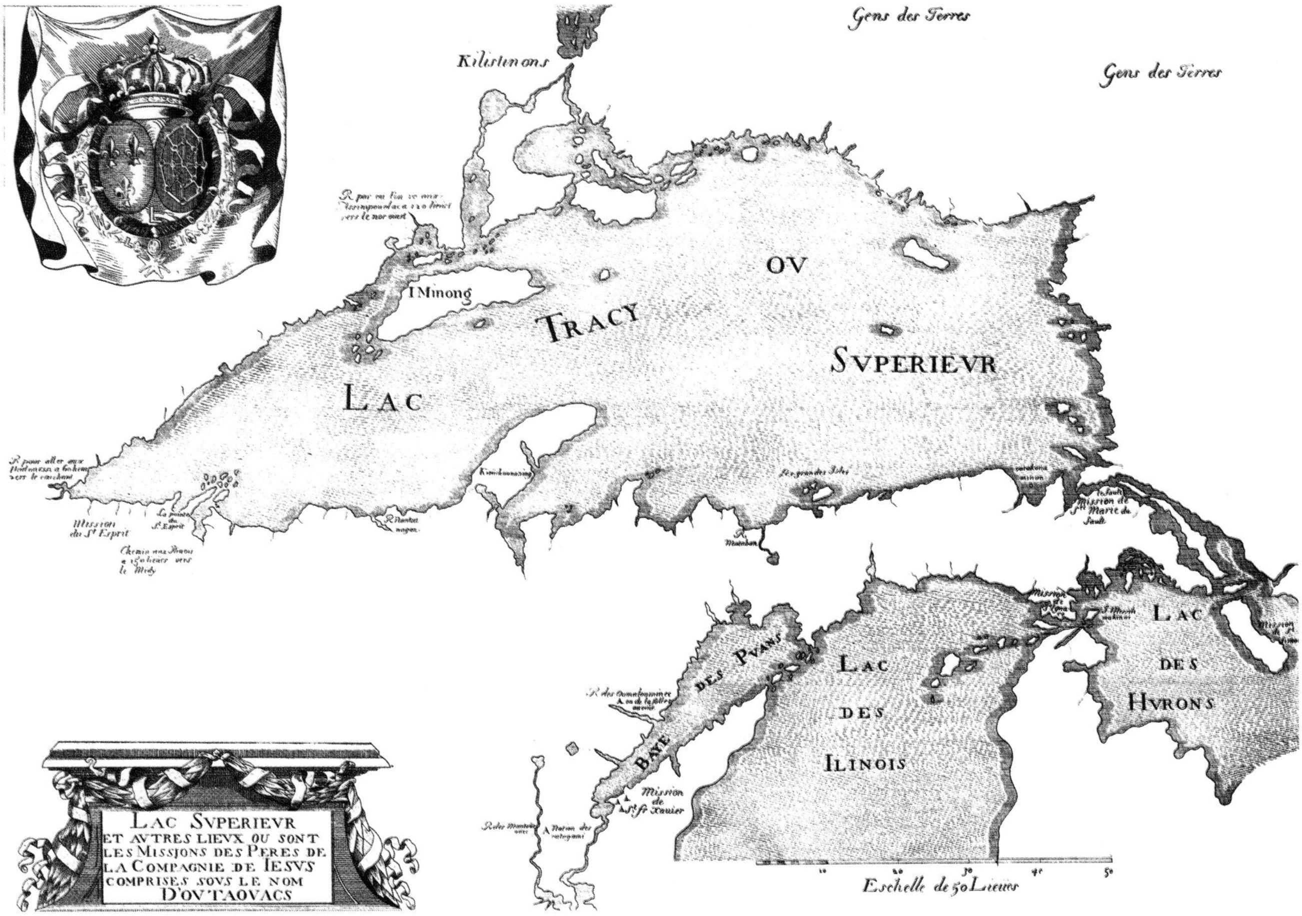

Gens des Terres

Kilistinons

Gens des Terres

LAC TRACY OV SVPERIEVR

I Minong

LAC DES ILINOIS

LAC DES HVRONS

BAYE DES PVANS

Mission du S.t Esprit

Mission de S.t Xavier

Mission de S.te Marie du Sault

Eschelle de 50 Lieues

LAC SVPERIEVR ET AVTRES LIEVX OU SONT LES MISSIONS DES PERES DE LA COMPAGNIE DE IESVS COMPRISES SOVS LE NOM D'OVTAOVACS

An 1883 facsimile of Lac Tracy ou Superieur, *a map made by Jesuit missionaries in 1670–1671 and published in Paris in 1672.* Lac Superieur *simply meant "the upper lake."* NEMHC RR MP D2

The Native Americans of the Great Lakes region also often used birch bark to construct dwellings. During much of the time that the French and British contested control of the fur trade in the Western Great Lakes region, the country west of Lake Superior was in ferment because of tribal conflict. In the eighteenth century the Anishinabe and Lakota clashed over possession of what today is Minnesota. The Anishinabe bested the Lakota in several pitched battles during the 1700s and drove their hereditary enemy to the short-grass prairie to the west. Zimmerman, NEMHC S2386 B34f4

Stephen Bungo (also spelled Bonga, in some accounts) was among the first African-Americans to live in what is now Minnesota. Bungo, who was reportedly born in 1799, was a freed man who claimed to have served a British officer during the War of 1812. He lived at Fond du Lac for most of his ninety years. As a half African-American and half Ojibwe man, Bungo always called himself "the first white man at the Head of the Lakes." NEMHC Portrait File S2386

The 1763 Treaty of Paris transferred all of the fur-trading rights and privileges from Montreal merchants to British citizens. An increasing number of pelts immediately began to flow through the Hudson's Bay Company factories to London instead of Paris. But the British and the Hudson's Bay Company enjoyed their monopoly for all of two decades.

A New Nation

St. Louis County owes its existence as part of the future state of Minnesota and the United States to Benjamin Franklin. The portly Philadelphia inventor, statesman, and ambassador joined John Jay, later the U.S. secretary of state; John Adams, the second president of the United States; Henry Laurens of South Carolina; and William Temple as the American committee to negotiate the 1783 Peace of Paris that would end the American Revolutionary War. By that point, Great Britain simply wanted to shed its rebellious colonies. The British peace commissioners also wanted to preserve as much of a commercial trading advantage as it could for the colonies it was keeping: Upper Canada, Lower Canada, and Prince Rupert's Land.

Franklin and the U.S. peace commissioners wanted to ensure that the new nation they were creating did not get shortchanged when the final treaty boundaries were drawn. For the most part, the task of the peace commissioners from both sides was fairly simple when it came to drawing the boundaries of the new nation and the British possessions to the north, at least east of the Great Lakes. Contention was created when the commissioners attempted to draw a boundary west of the Great Lakes, and that was frankly because the maps in use at the time were none too accurate.

Both sides used a 1755 map drawn by explorer and botanist John Mitchell, which led to much confusion in the treaty negotiations. Mitchell's map showed two lakes north and west of Lake Superior that flowed into Lake Superior via what he dubbed the "Long Lake," which was in reality the Pigeon River. More importantly, Mitchell showed the Mississippi River west and north of present-day Lake of the Woods. In effect, Mitchell had confused the Mississippi River, which rises near Bemidji, Minnesota, one hundred miles south of Lake of the Woods, with the Red River of the North, which flows north into Lake Winnipeg.

As the peace commissioners studied the maps, they naturally proposed boundaries that would give their client states the largest amount of territory. Franklin and the American commissioners suggested following the course of rivers and proposed a boundary west of Lake Superior that would run along the Rainy River to the "northwest angle" of Lake of the

By the 1820s, Fond du Lac was a thriving village populated by American, French, and English-Canadian fur traders, as well as a sizable contingent of Native Americans. When Henry Rowe Schoolcraft was appointed U.S. Indian Agent for the Lake Superior Region in 1822, he reported that as many as 375 people lived in the area immediately surrounding the post. NEMHC S2386 B13f1

By the 1850s, European settlement at the Head of the Lakes was concentrated on the east bank of the St. Louis River in the state of Wisconsin and on the sandbar that the first settlers would call Minnesota Point. The American Fur Company trading post upriver at Fond du Lac quickly fell into disrepair and was torn down in the 1890s. NEMHC S2386, Box 13f1

Woods, and then west to the Mississippi River. The boundary of the new nation would then follow the Mississippi south to the Spanish possessions at St. Louis and New Orleans.

The British rejected the idea and suggested drawing the boundary along the forty-seventh parallel west of Lake Superior to the Mississippi. That suggestion would have created the boundary just north of the fur post of Fond du Lac, and then west to the Mississippi. Franklin and the Americans countered with a proposal to adjust the boundary in Lower Canada between the frontier fort of Detroit and Niagara Falls, but the British commissioners objected to the loss of so much prime agricultural land in what is today southern Ontario.[10]

C. C. Rosenkranz's 1919 oil painting imagining the meeting between the Sieur du Lhut and the local inhabitants at Little Portage, Minnesota Point, on June 27, 1679. St. Louis County Historical Society Museum

Franklin guided the negotiations back to the initial proposal. In the late summer of 1783, both sides agreed on drawing the boundary west from the Pigeon River through Rainy Lake and Lake of the Woods to where it crossed the supposed course of the Mississippi River. Although neither the United States nor the British commissioners realized it at the time, the boundary that Franklin suggested and was eventually agreed to gave the United States possession of three of the most productive iron ore ranges in the world. The 1783 treaty negotiation also laid the groundwork for the geopolitical creation of St. Louis County nearly seventy-five years later.

The final boundary wouldn't be completely settled until 1842. Thomas Jefferson's 1804 purchase of the Louisiana Territory from Napoleon created a vast new territory west of the Mississippi that required drawing boundaries between it and British Canada to the north. An 1818 boundary commission employed David Thompson, a former Hudson's Bay Company surveyor, to survey the disputed boundary. Thompson had long argued that the source of the Mississippi was far south of where the 1783 boundary put it. In 1832, Giacomo Beltrami discovered the source of the Mississippi in a meadow south of present-day Bemidji and named it with a Latin anagram—*Itasca*—"truly the head."

Finally, in 1842, the Webster-Ashburton Treaty accepted the Pigeon River route west of Lake Superior and drew a line south from the northwest angle of Lake of the Woods to the forty-ninth parallel.[11] The boundary then followed the forty-ninth parallel two thousand miles to the Pacific Ocean.

An American Fur Empire

While the United States and Great Britain were painfully negotiating a boundary west of Lake Superior during the sixty years between 1783 and 1842, the Hudson's Bay Company and various American and Canadian fur companies were competing for control of the fur trade in the western Great Lakes.[12]

The North West Company was a joint stock firm formed by Montreal merchants and *voyageurs* during the American Revolution. North West Company agents spread across the land north and west of the Great Lakes and provided fierce competition to the Hudson's Bay Company. By the time the Treaty of Paris was inked in 1783, the Hudson's Bay Company had moved the focus of its operations to the Canadian Rockies.

By 1800, the North West Company dominated the fur trade of the Great Lakes and its hinterlands. Its more than one thousand traders and clerks buying pelts and sending them back along the traditional canoe routes to Montreal included Jean Baptiste Perrault, who built a trading post at Fond du Lac. Perrault also built smaller posts upriver on the St. Louis, as well as at Leech Lake, Pine Lake, and Ottertail Lake.[13]

More competition came in 1795 when disaffected partners of the North West Company formed the XY Company.[14] With traders from three major fur outfits traversing American and British territory west of the Great Lakes, competition was becoming a way of life for both Native Americans and Europeans alike.

As the market for beaver-skin hats inevitably tailed off in the first half of the nineteenth century, the competition in Minnesota got sharper still. In 1808, John Jacob Astor formed the American Fur Company to compete with the Canadians operating south of the international boundary. American Fur Company traders took over the North West Company's posts at Fond du Lac and La Pointe on Lake Superior's Madeline Island. The U.S. traders controlled the fur trade west to the Rockies south of the forty-ninth parallel from a major fur post at Mackinac Island.

Business was disrupted during the War of 1812, and Astor sought legislation after the war that would essentially bar Canadian traders from American soil. Congress obliged in 1816, and five years later, the North West Company admitted defeat at the hands of Astor and the Americans and merged with the hated Hudson's Bay Company.[15] For the next two decades, the American Fur Company reigned supreme in the U.S. fur trade.

A quarter century after founding the American Fur Company, Astor retired in 1833 to a life of leisure and philanthropy. The industry that had provided the economic underpinning of the North American heartland for more than two hundred years was rapidly drawing to a close. Fashionable gentlemen in London, Paris, and St. Petersburg were beginning to sport silk top hats with their evening wear. Beaver hats were suddenly so passé.[16]

The year following Astor's retirement, the American Fur Company reorganized and closed nearly half of its trading posts.[17] Nothing the company did, however, stemmed the flow of red ink caused by the change in fashion. In 1842, the American Fur Company became one of the largest firms in the short history of the American republic to seek bankruptcy protection.

As the two-hundred-year fur trade of the Western Great Lakes inexorably drew to a close, new opportunities beckoned. Within a decade of the 1842 collapse of the American Fur Company, the company's trading post at Fond du Lac would be supplanted by a new settlement on the shore of nearby Lake Superior.

Detail of a five-foot by eight-foot art glass window displayed in the St. Louis County Heritage and Art Center in downtown Duluth depicting Daniel Greysolon at Minnesota Point in the summer of 1679 greeting an unidentified Ojibwe Indian. The window was designed by Axel Burgholz of Duluth's St. Germain Brothers Glass Company in 1928 for the first floor of Duluth's new City Hall building. St. Louis County Heritage and Arts Center

1856 Interlude

Eighteen fifty-six was a watershed year for St. Louis County. Territorial legislators, meeting in St. Paul, recognized the reality of the growing population in Minnesota's northeastern corner when they formally organized St. Louis County and Lake County out of Itasca County on March 1, 1856. The wind-down of fur trading in the 1840s had resulted in a thinning out of northern Minnesota. When the Minnesota territory took its first census in 1850, only ninety-seven Yankees—literally all who had come from the East—were recorded in Itasca County, which at the time consisted of most of the northeastern quadrant of the state.[1]

But three events outside the Minnesota territory in the late 1840s and early 1850s created a jump in the number of inhabitants along the North Shore of Lake Superior and in adjacent northwestern Wisconsin. The discovery of commercial deposits of iron ore near present-day Negaunee in Michigan's Upper Peninsula in the mid-1840s and commercial deposits of high-grade copper at Eagle River on the state's Keweenaw Peninsula several years later created a mining boom at the western end of Lake Superior.

In the fall of 1844, surveyors running a base line south of Marquette, Michigan, recorded major variations of the magnetic needle in their compass. The surveyor instructed the party to "look around and see what you can find," and the assistants soon were tearing off chunks of iron ore from the surrounding outcrops.[2]

Already in that summer of 1844, prospectors were flocking to Sault Ste. Marie to take passage on the propellor *Independence* and the sidewheeler *Julia Palmer,* both of which had been hauled overland past the rapids of the St. Mary's River, for the copper discoveries awaiting on the rock-bound Keweenaw Peninsula.[3] Douglass Houghton, the state's charismatic geologist, had first identified the potential of copper development on the Keweenaw in the early 1840s. Houghton, who gave his name to the major town in the Copper Country mining district, wouldn't live to see his suspicions of copper riches confirmed. He died in a boating accident off Eagle Harbor in October 1845.[4]

Houghton and the U.S. surveyor on Teal Lake just east of the Keweenaw had set into motion the first major U.S. mining boom. Cleveland entrepreneurs formed syndicates to exploit the iron ore resources south of the port city of Marquette, and by the early 1850s, a number of charcoal furnaces had been erected in the region to smelt the raw ore into a more usable form of low-grade pig iron.[5]

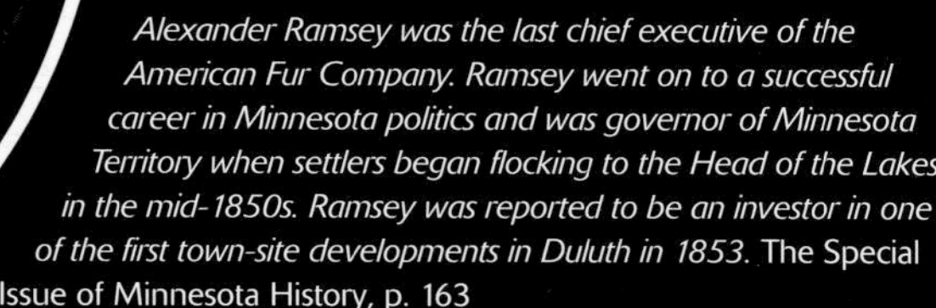

Alexander Ramsey was the last chief executive of the American Fur Company. Ramsey went on to a successful career in Minnesota politics and was governor of Minnesota Territory when settlers began flocking to the Head of the Lakes in the mid-1850s. Ramsey was reported to be an investor in one of the first town-site developments in Duluth in 1853. The Special Issue of Minnesota History, p. 163

By 1846, the Cliff Mine near Eagle River was producing more than a half-million pounds of easily smelted native copper. Three years later, in 1849, the Cliff Mine paid an initial dividend to its Pittsburgh investors, establishing a reputation for profitability for Michigan's copper industry that would endure for a century.[6]

Edmund F. Ely lived in the Lake Superior wilderness country for more than thirty years during the middle of the nineteenth century. Born and raised in Massachusetts, the twenty-three-year-old Ely accepted an assignment from the American Board of Foreign Missions to minister to the Anishinabe in 1832. For much of the next three decades, Ely served missions at Fond du Lac and Pokegema in Minnesota. From 1856 to 1862, he ministered in the growing community of Oneota. Ely left the Head of the Lakes during the Civil War for St. Paul and later moved to California, where he died in 1892. NEMHC Portrait File

The Treaty of La Pointe

In order for the new nation to create a land ownership society in the Lake Superior country, what was euphemistically referred to at the time as "the Indian problem" had to be addressed. The Anishinabe were perhaps the most accommodating of all the Native American communities the new republic would encounter in the first half of the nineteenth century. By the 1820s, the Lake Superior Anishinabe had been trading with Europeans for nearly two hundred years.

Unlike their more warlike neighbors to the south, the Sac and Fox, the Anishinabe maintained amicable relations with the Indian agents, traders, and settlers they interacted with during the 1820s and 1830s. But for settlement to take hold in the Upper Great Lakes required a transfer of land ownership from the Native American community to citizens of the new republic. U.S. Indian agents began laying the groundwork for that ownership transfer in 1837, when they convinced a group of Anishinabe chiefs to sign over title to a large chunk of northwestern and northern Wisconsin, as well as a substantial portion of what was to become east-central Minnesota.

In 1842, Anishinabe bands ceded most of the western half of the Upper Peninsula of Michigan and adjacent territory in Wisconsin from north-west of Green Bay to Lake Superior.[7] The Anishinabe, whose concept of land ownership was more in character with lifetime stewardship than actual legal possession of the land, were content with the six reserves that the federal government delineated for them in the ceded area of the 1842 treaty.

The discovery of iron and copper in the Upper Peninsula within several years of the 1842 treaty put immense pressure on the government to open up additional land in the Lake Superior basin. In 1854, the U.S. Commissioner of Indian Affairs asked all of the bands of Mississippi River and Lake Superior Anishinabe to gather at La Pointe on Madeline Island to discuss further cessions. For two weeks in September, the U.S. commissioners and eighty-five representatives of the Anishinabe bands dickered over the terms of the proposed cession, the largest so far.

On September 30, 1854, the parties agreed to a land transfer that essentially comprised the entire Arrowhead Region of what was soon to become the state of Minnesota. After trying and sometime contentious negotiations, the land cession included all of the land east of a line from the Snake to the Savanna and St. Louis to the Vermilion and Rainy Rivers.[8]

The 1854 cession opened up the North Shore of Lake Superior to white settlement and legitimized the tiny communities at the Head of Lake Superior. Duluth, situated at the foot of the massive ridge overlooking the western end of the Lake, had been settled just that summer. Superior City, across the bay in Wisconsin, was little more than a year old in 1854. The Treaty of La Pointe also took in substantially all of the land that would be assigned to St. Louis County just two years later.

The Anishinabe received a payment of about $1 million and title to reservations in the ceded territory at Fond du Lac, Nett Lake, Grand Portage, and Mille Lacs. The payment of about $1 million was in small amounts over a twenty-year period. The Fond du Lac band ended up controlling more than one-hundred thousand acres above Knife Portage on the St. Louis.[9] With the land question settled by 1854, the rapid settlement of northeastern Minnesota was restricted only by logistical problems.

Locking through the Soo

In the mid-1850s, the settlement of Minnesota Territory was accelerating. Rail lines from neighboring Wisconsin reached the Mississippi River by 1855, and settlers could take steamboats north on the Mississippi from Prairie du Chien and La Crosse to the Falls of St. Anthony at present-day Minneapolis.[10]

The Reverend Edmund Ely sketched this map of the fur post at Fond du Lac in 1837. Note he position of the Anishinabe village at the ip of an island in the middle of the St. Louis River. Edmund Ely, NEMHC SA File

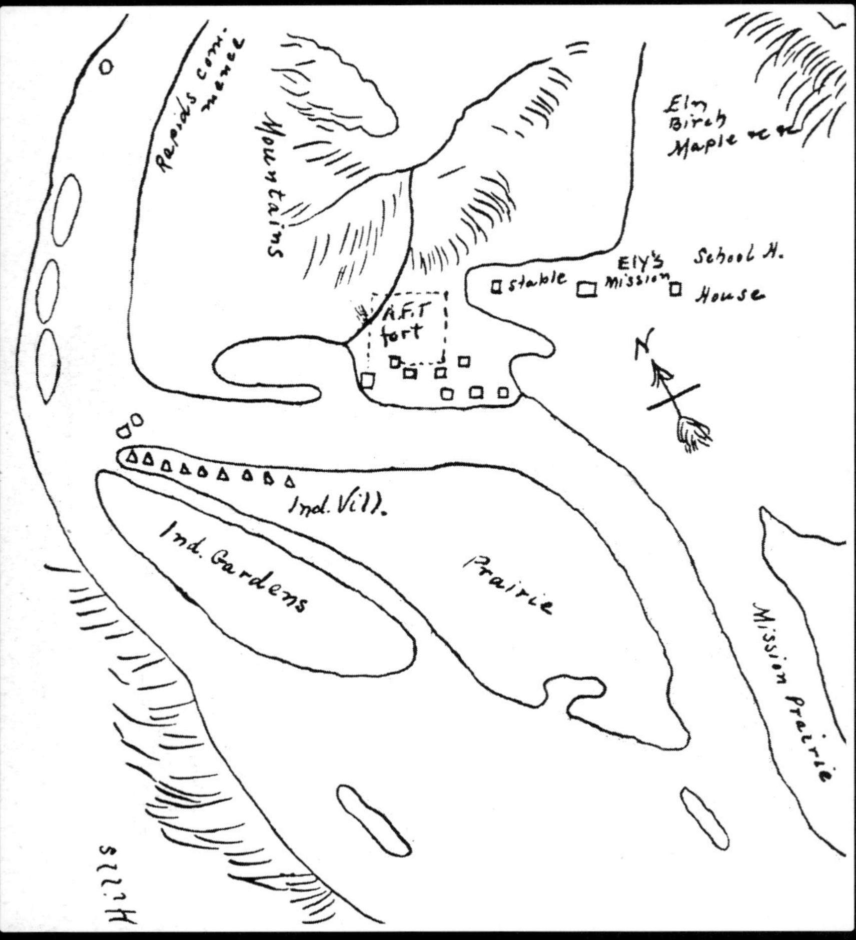

Congress in the early 1850s had commissioned the construction of a military road from St. Paul to Lake Superior. The 150-mile route, which actually terminated on the Wisconsin side of the border at Superior City, crossed swamps and forests and tested the endurance of even the hardiest settler. The military road was a morass during the spring and wet weather. Its passage was easiest by dog sled in the dead of winter.

As late as 1869, before the railroad reached the Head of the Lakes, travelers destined for Duluth from St. Paul often opted to go the long way around rather than hazard the military road. The Reverend Harvey Webb, who accepted a call from the Methodists at the Twin Ports, set out from the Falls of St. Anthony for Duluth with his family and seventeen hundred pounds of household goods on October 15, 1869. The Webbs took the railroad east to Milwaukee, where they changed trains for Green Bay and Marquette. At Marquette, they booked passage aboard a steamer for Superior City. They arrived at their parsonage on October 31, 1869.[11]

The drive north from the Twin Cities that nobody even thinks about today and takes a little more than two hours on Interstate 35 required sixteen days to accomplish via rail and steamer in the early days of St. Louis County.

The Webb family's decision to travel for at least part of their journey on Lake Superior demonstrated that a far easier way to get to northeastern Minnesota in the mid-nineteenth century was by water rather than by land. As the mining booms erupted on the Upper Peninsula, Congress began to seriously study expediting the flow of commerce to the Upper Lakes. The major bottleneck for the sailing ships, sidewheelers, and propellors that had supplanted Montreal Canoes on the Great Lakes was the series of rapids that dropped the St. Mary's River twenty-one feet in a torrent of white water between Lakes Superior and Huron at Sault Ste. Marie.

As early as the 1790s, North West Company traders had built a small lock and canal at the Soo to allow the passage of *Canot du Maitre* (Montreal Canoes) and *bateaux* without having to unload them for portage around the rapids. American troops had destroyed the lock and canal during the War of 1812, and it wasn't until the discovery of iron ore and copper on the Upper Peninsula that Congress and the State of Michigan began seriously examining the feasibility of a major lock and canal system at the Soo.

Edmund Ely's helpmate for most of his years in the Lake Superior country was Catharine Goulais Ely. The daughter of a French-Canadian father and Anishinabe mother, Catharine Ely was educated in the mission schools on Mackinac Island and married Edmund Ely at La Pointe in 1835. Ely was stationed at Fond du Lac at the time, and their daughter, Mary, was born at the Head of the Lakes in the spring of 1836. Catharine Ely would bear twelve more children between then and 1863. Only six lived past childhood. NEMHC Portrait File

Finally, in 1852, Congress sent legislation to President Millard Fillmore that set aside more than three-quarters-of-a-million acres of Upper Peninsula land for payment to a private company for construction of a lock and canal system on the St. Mary's River. Some in Congress thought the Soo Locks pork barrel politics at their most pungent. Henry Clay of Kentucky, one of the great statesmen of the nineteenth century, had gone on record as early as 1840 opposing the locks as "a work beyond the remotest settlement of the United States, if not the moon."[12]

Senator Clay's objections aside, the Michigan Legislature also quickly provided incentives for completion of the mile-long canal and double lift locks and oversaw the submission of bids for the project. The St. Mary's Falls Ship Canal Company, a consortium of Eastern manufacturers well aware of the iron and copper potential of the Upper Peninsula, submitted the winning bid of $400,000.

The Ship Canal Company sent twenty-four-year-old Charles T. Harvey to the Soo in the late spring of 1853 with enough equipment to outfit four hundred men and a $50,000 line of credit. During the next two years, Harvey and his crews worked around the clock in mud, snow, and dust. They blasted through solid rock, dug a channel more than a mile long, and erected the huge gates that alternately helped fill and empty the locks. By the time the locks filled with water, Harvey and the Ship Canal Company were providing employment for more than two thousand laborers.[13]

The Lake Superior end of the canal began filling with water in mid-April 1855.

he locks at Sault Ste. Marie made possible both the rapid ettlement and commercial exploitation of the Upper Great akes. The locks worked by a simple principle still in use oday. By manipulating the sluice doors on either sides of he lock, an operator can raise a vessel upbound from Lake Huron the twenty-one feet to the level of Lake Superior. By reversing the process, a vessel can be gently lowered he twenty-one feet from Lake Superior to Lake Huron. The Soo Locks celebrated their 150th anniversary in he summer of 2005, one year before St. Louis County elebrated its 150th anniversary. NEMHC S2386

The opening of the Soo Locks in 1855 meant much more ommercial navigation on the Upper Lakes. Concurrent with the opening of the locks, the federal government ontracted for the construction of a lighthouse at Minnesota Point near the Superior Entry. The $15,000 lighthouse ncluded a Bardon Lens made in France and was illuminated by a kerosene lantern. The federal appropriation for he lighthouse was the first of millions of dollars the U.S. government would expend to improve the harbor at he Twin Ports during the next 150 years. NEMHC S2386 B30f19

shipload of iron ore loaded at Marquette.[14] With the sluice gates open on the Lake Superior end of the canal, the floodgates were about to open for the Lake Superior country.

The First Settlement

George R. Stuntz, a Pennsylvania native, came to the western end of Lake Superior in 1852 as a surveyor. He helped map the boundary between Minnesota Territory and the State of Wisconsin, as well as the site of Superior City. Then thirty-two years old, he was ready to settle down in one place. The next year, he returned to the Head of the Lakes and built a warehouse and dock on Minnesota Point near the Superior Entry. Besides Stuntz, who wintered over in 1853–1854, the only other U.S. citizens on the Minnesota side of the bay were Reuben Carlton, the blacksmith at Fond du Lac, and George Nettleton, a licensed trader with the Native American community.[15]

George Nettleton is credited with being the incorporator of the town site of Duluth. Nettleton and his brother, William, arrived at the Head of the Lakes in about 1853 from the Crow Wing Agency in Minnesota Territory. The Nettletons were licensed traders with the Anishinabe, and Nettleton told his wife that there was bound to be a city someday at the Head of the Lakes. Eye of the Northwest, p. 9

Far more settlers lived on the Wisconsin side of the bay, primarily because the land on the Minnesota side was still owned by the Anishinabe. Stuntz and Nettleton lived on the Minnesota side only because they were licensed traders.

Superior City, meanwhile, boasted two Presbyterian churches with a total of twenty-nine members between them.[16] In the earliest days of settlement, the local ministers spent as much time fighting "demon rum" as they did preaching. In October 1856, the Reverend James Peet confided to his diary that "the *Lady Elgin* came in; brought heavy cargo of provisions and enough whiskey to keep the whole town drunk all winter."[17] The Wisconsin settlement even had a weekly newspaper to help catch residents up on the news of the day.

The mining fever that swept the Lower Great Lakes in the 1850s, coupled with the Treaty of La Pointe in 1854 and the opening of the Soo Locks the next summer, provided all the incentive hundreds of settlers needed to move to the far distant Head of the Lakes. When the federal government built the Minnesota Point lighthouse in 1855, it signaled that the new territory was ripe for development.

In 1855 and 1856, town sites popped up along the lakeshore on the Minnesota side at Beaver Bay, Oneota, Duluth, Portland, and Buchanan. The Minnesota Territorial Legislature's 1856 decision to carve out a huge county in northeastern Minnesota was further affirmation that the land west and north of Lake Superior would soon be bustling with industry.

George R. Stuntz first saw the Head of the Lakes in 1852 when he arrived as part of a federal government survey party. A Pennsylvania native, Stuntz was thirty-two at the time and he would spend the next fifty years of his life in Duluth. Stuntz wrote in 1892 that as he first stood at Minnesota Point in 1852, "I saw the advantages of this Point as clearly then as I do now. I went away to make a report and returned the next spring and came for good. I saw as surely then as I do now that this was the heart of the continent commercially, and so I drove my stakes." Gaylord, NEMHC S3766 B1f31

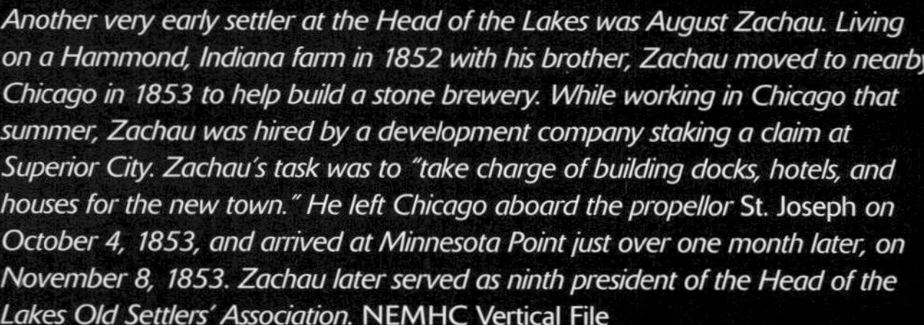

Another very early settler at the Head of the Lakes was August Zachau. Living on a Hammond, Indiana farm in 1852 with his brother, Zachau moved to nearby Chicago in 1853 to help build a stone brewery. While working in Chicago that summer, Zachau was hired by a development company staking a claim at Superior City. Zachau's task was to "take charge of building docks, hotels, and houses for the new town." He left Chicago aboard the propellor St. Joseph *on October 4, 1853, and arrived at Minnesota Point just over one month later, on November 8, 1853. Zachau later served as ninth president of the Head of the Lakes Old Settlers' Association.* NEMHC Vertical File

Colonel Joshua B. Culver was one of the first merchants at the Head of the Lakes. Shortly after the Treaty of La Pointe was ratified by Congress in 1855, he was associated with the Nettletons in platting the town site of Duluth and opened a mercantile store on Lake Avenue that same year. Colonel Culver also had an interest in a sawmill on Lake Avenue and was involved with platting communities on the North Shore and up the St. Louis River in 1856 and 1857. Culver was named the first Clerk of the District Court when St. Louis County was formed in 1856. NEMHC Portrait File

Colonel Hiram Hayes was perhaps the first attorney at the Head of the Lakes. He arrived in Superior City in 1854 at the age of twenty-two and remained a resident of Douglas County, Wisconsin, for the remainder of his long life, dying in Superior in 1918. Hayes, like most of the other early residents of the area, was an investor in the town-site developments that sprouted on both sides of Superior Bay. In 1856, he was one of the investors in the town site of Burlington Bay on the North Shore. Although Hayes lived in Superior from 1854 on, he once reportedly served as a judge of probate in nearby Duluth. NEMHC Portrait File

The first newspaper at the Head of the Lakes was the Superior Chronicle, *established in the spring of 1855 by Washington Ashton and John C. Wise and housed in a log cabin. The two Virginians had previously worked for the* Congressional Globe *in the nation's capital. They published their weekly newspaper as "the voice of one crying in the wilderness to proclaim to the world the Gospel of city building and fortune making."* NEMHC S2386 B20f2

Crossroads of a Continent

chapter two

Because St. Louis County has depended on the extraction of natural resources for most of its history, the area has always been susceptible to economic dislocations experienced elsewhere in the North American community. One of the more piquant observations about the county's economic health is summed up in the statement, "When the steel industry catches cold, the Range comes down with pneumonia."

The wisdom of that essential truth was demonstrated early in the county's history. In 1857, one year after the Territorial Legislature formed the county, the failure of a bank in Ohio quickly spread to the nation's financial industry.

Americans of the nineteenth century called financial disruptions *panics*, and the terminology is descriptive of what happens when ordinary Americans lose faith in the financial system. Americans of the twentieth century used the more euphemistic terms *depression* and *recession* to describe the loss of confidence in banks and money, but *panic* certainly takes into account the psychological as well as economic ramifications of downturns in the national economy.

Minnesota Territory was particularly affected by the Panic of 1857, because the soon-to-be state's economy rested on three sectors that were especially impacted by the downturn: farming, railroads, and land speculation. The territory, which would achieve statehood in 1858, was in the midst of a railroad-building boom.

Because St. Louis County has depended on the extraction of natural resources for…

Settlers from Germany and Scandinavia were flocking into the southern half of the territory, attracted by some of the richest farmland in the Great Lakes. They harvested a bumper crop of wheat in 1856 and 1857, but the end of the Crimean War in 1856 created a worldwide wheat glut when farmers in Southern Russia re-entered the

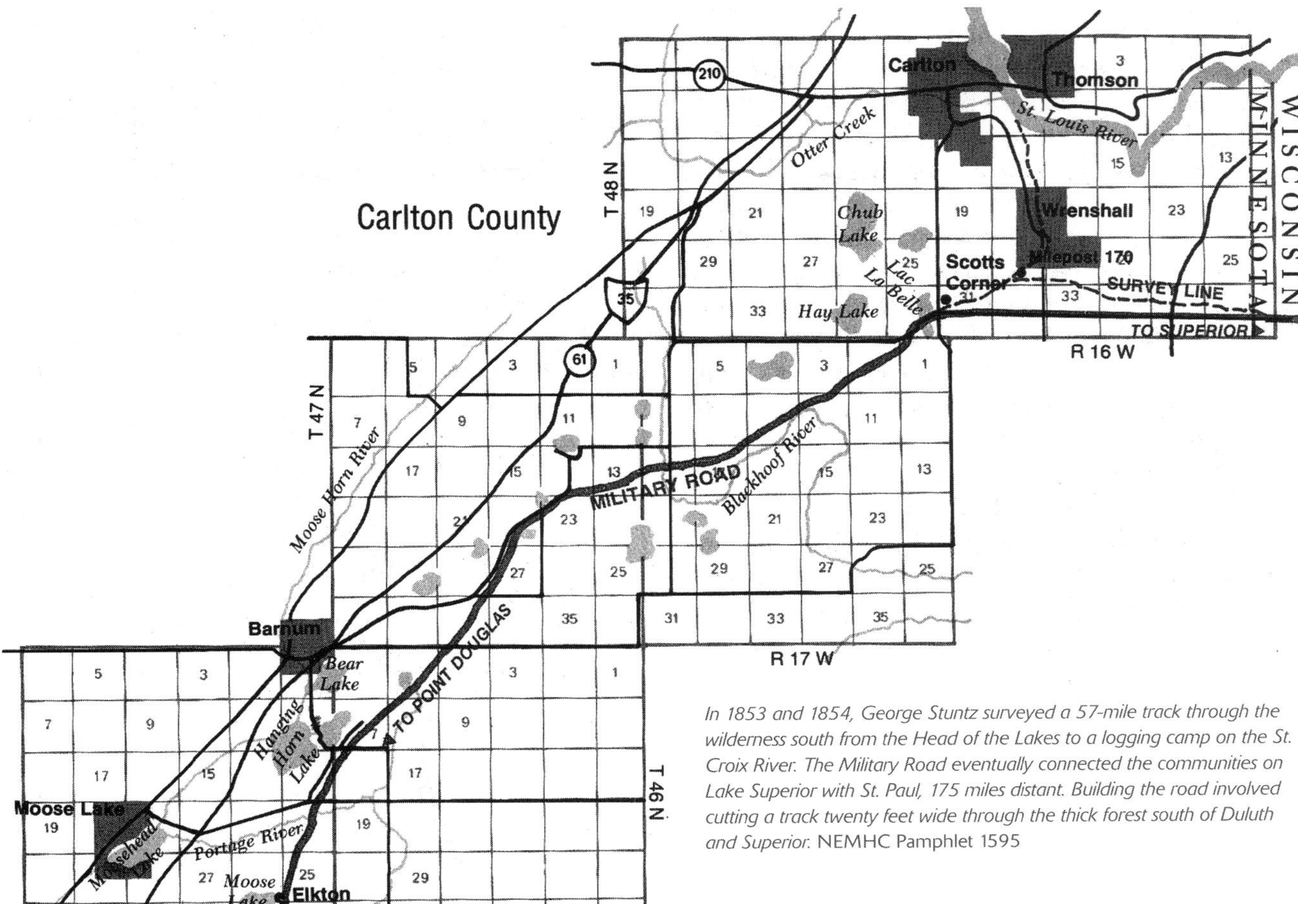

In 1853 and 1854, George Stuntz surveyed a 57-mile track through the wilderness south from the Head of the Lakes to a logging camp on the St. Croix River. The Military Road eventually connected the communities on Lake Superior with St. Paul, 175 miles distant. Building the road involved cutting a track twenty feet wide through the thick forest south of Duluth and Superior. NEMHC Pamphlet 1595

marketplace.[1] Prices plummeted, and Minnesota farmers were unable to pay their debts. Banks failed, and railroads building across the territory began to shut down.

Duluth, Portland, Oneota, and the other town sites along the harbor at the Head of the Lakes might have survived the Panic of 1857. In the territorial capital of St. Paul, there was talk of building a rail line north to Lake Superior, but no construction had actually begun. And there were no wheat farmers in St. Louis County to be affected by a deluge of Crimean grain flooding the market.

What there was at the Head of the Lakes was rampant land speculation. When the Panic hit the communities at the western end of Lake Superior, the devastation was swift and complete. Land values were in free fall by early summer. The bulk of the vessels calling on the Twin Ports during the summer of 1857 were swamped with recent residents leaving the area for more settled regions on the Lower Great Lakes. One writer noted that "the lake steamers were overcrowded, and the military road to St. Paul had more south-bound pedestrians than Coxey's army ever numbered."[2]

The Head of the Lakes had not yet recorded its first census, but observers estimated the population at between two and three thousand in early 1857, most living on the Superior side of the harbor. By the end of the year, fully three-quarters of the residents had packed up and left. Communities on the North Shore, including Agate Bay, Silver Creek, Encampment, and Grand Marais, were literally abandoned by 1858.[3]

It would take St. Louis County and the town sites near the county seat of Duluth nearly a dozen years to recover from the ravages of the Panic of 1857. The 1860 federal census reported only 80 non-Anishinabe residents on Minnesota Point.[4] The census-takers reported 406 people in the entire county, and many of them were Anishinabe. By that time, St. Louis County had already spun off Carlton County to the south and west,

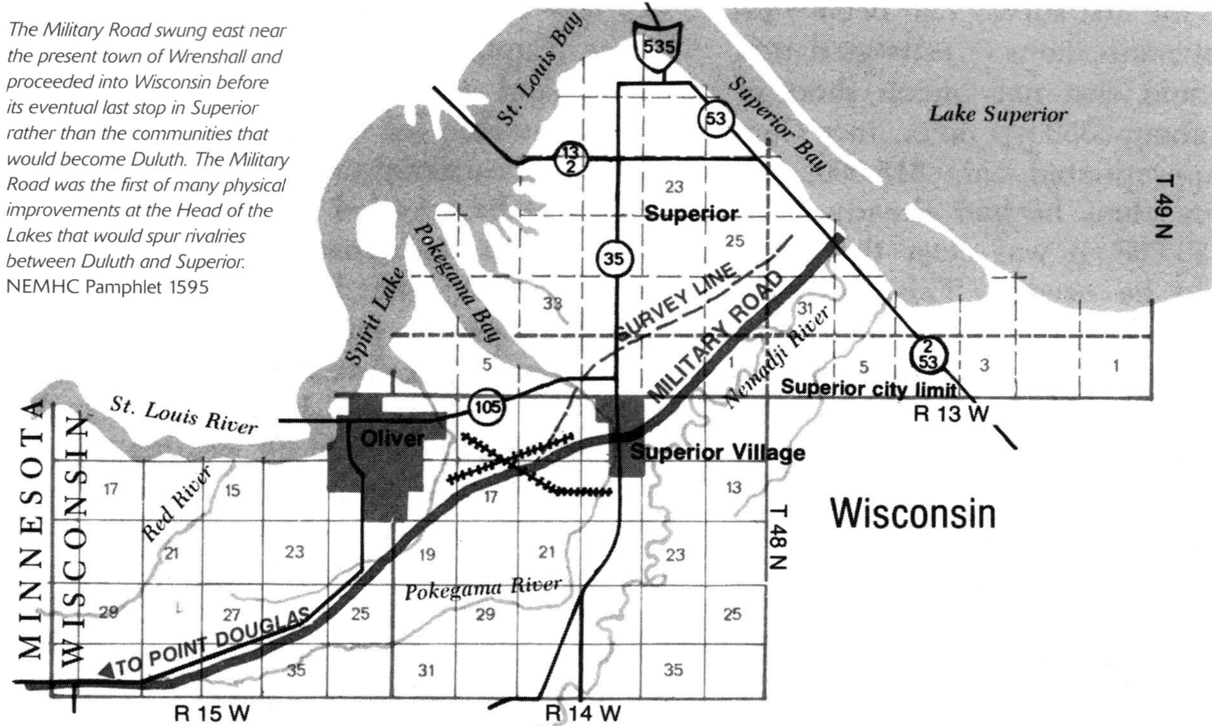

The Military Road swung east near the present town of Wrenshall and proceeded into Wisconsin before its eventual last stop in Superior rather than the communities that would become Duluth. The Military Road was the first of many physical improvements at the Head of the Lakes that would spur rivalries between Duluth and Superior. NEMHC Pamphlet 1595

and Lake County up the North Shore. Carlton County in 1860 had a population of a mere 51 people, and Lake County's population was 248.

Alfred Merritt had arrived at the Head of the Lakes with his parents and brothers from Ohio in the fall of 1856. The Merritts stayed after the Panic, and Alfred later recalled that "as a boy, I knew every man and woman on this side of the state line."[5]

The County Emerges

As a county in the new state of Minnesota, St. Louis County had to create a county administration during the low population years of the late 1850s and 1860s. The first Board of County Commissioners met in Duluth in January 1858 to draw a grand and petit jury. Four townships were laid out—Oneota, Fond du Lac, Duluth, and Carp River—and school districts were organized.

The County Board also took bids for construction of the first road in the county, which would eventually run from Fond du Lac to Oneota through Duluth and along the North Shore past Lester River to the mouth of the Knife River. Through the 1860s, however, county residents typically took the water route if venturing to Knife River and beyond.[6]

The growth of copper mining in Michigan's Upper Peninsula during the Civil War years fueled several copper mining developments in the 1860s, both at French River on the North Shore and along the South Shore east of Superior City. But the event that triggered renewed interest in settlement in northeastern Minnesota was the 1866 account by the Minnesota state geologist of the possibility of gold in the far northern reaches of St. Louis County.

Henry H. Eames had mapped the county's northern quadrant in 1865–1866 and reported back to St. Paul that commercial

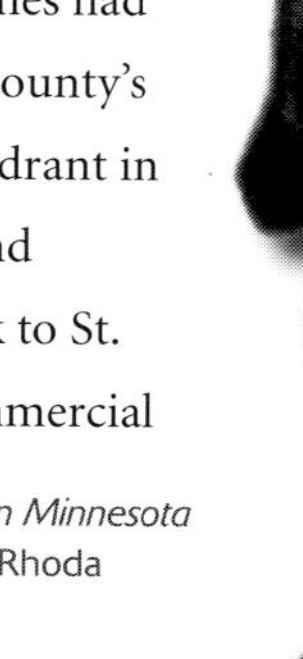

Henry H. Sibley was a contemporary of Henry M. Rice and another giant of early state politics. When Minnesota went to the polls in 1858 in the state's first election, residents elected Sibley the state's first governor. Rhoda Gilman, The Story of Minnesota's Past, p. 94

Henry M. Rice was one of the giant figures in the history of Minnesota statehood. In 1855, Rice, then the Minnesota Territory's delegate to Congress, secured an appropriation of $20,000 to build bridges and maintain the Military Road. Special Issue, Minnesota History, Winter 1998, p. 167

quality gold could be found in the hills overlooking Lake Vermilion.[7] Within weeks, hundreds of men swarmed north from Duluth, Portland, and Oneota seeking their fortunes in the rumored gold fields. The gold rush was short-lived.

George Stuntz, who had been among the first to trek north to investigate the stories of gold, soon returned to Duluth with news that the reports of gold were false. But Stuntz did confirm Eames' overlooked information that the hills surrounding Lake Vermilion were underlain with iron, with deposits in some cases upward of sixty feet thick and specimens assaying as high as 80 percent iron.[8]

The gold rush did help open up the north end of the sprawling county. A post office, general store, sawmill, and several saloons were hacked out of the wilderness at the town site of Winston, the first settlement in the county outside the cluster of communities on the lakeshore sixty miles south. Stuntz soon after contracted with the U.S. War Department to build a road from the Head of the Lakes to Lake Vermilion. The Vermilion Trail, completed in 1868–1869, connected Duluth with Lake Vermilion and the Bois Fort Indian Reservation at Nett Lake.[9]

Enter Mr. Cooke

Roads and the potential mineral wealth of northeastern Minnesota were on the minds of many in the years just after the Civil War. The gold prospectors who rushed north in the wake of Eames' report in 1866 dreamed of riches for the taking, but capitalists in St. Paul, Chicago, New York, and Philadelphia schemed of fortunes for the asking.

Eastern capitalists controlling the destiny of northeastern Minnesota with little thought of the local people—whose labor provided the capitalists' daily bread—would be a recurring theme in St. Louis County politics for the next century. And the first great capitalist to attempt to expand his already sizable fortune in St. Louis County was the model for all the rest.

Jay Cooke was perhaps one of the best-known businessmen in America at the close of the Civil War. A native of the Lake Erie Firelands, Cooke had worked for Philadelphia banking firms during the 1850s. In 1861, as Confederate artillery pounded Fort Sumter to signal the onset of the Civil War, Cooke founded his own investment banking firm, Jay Cooke and Company. During the next four years, the Philadelphia firm almost singlehandedly financed Mr. Lincoln's armies, underwriting more than $350 million in war bonds by 1865.[10]

The war made Cooke a wealthy man, but by no means rich, and he soon began seeking investment opportunities on the nation's expanding western frontier. The Head of the Lakes was among the first places he looked.

In 1864, the federal government made available 1.5 million acres of land along the right-of-way of a proposed railroad from St. Paul north to Lake Superior. The homestead provisions of the Morrill Act meant that railroad companies could accept federal government land for building a rail line and finance the construction of the rails and ties with the sale of land to settlers.

Cooke was intimately familiar with the legislation, and in 1866, he and his partner, William Moorhead, purchased more than forty thousand acres of land near Duluth. Cooke and Moorhead also bought several waterpower sites along the St. Louis River near what would later become Jay Cooke State Park and the Thomson and Fond du Lac Dams.

Two years later, in 1868, Jay Cooke and Company purchased bonds in the Lake Superior and Mississippi Railroad, which at the time was nearing the completion of construction north from St. Paul.[11]

William H. Hearding, born in England in 1826, was employed by the U.S. Lake Survey for forty years, from 1851 to 1891. In 1861, Billy Hearding led a crew of surveyors to the Head of the Lakes to complete the first survey of the Duluth-Superior harbor. Hearding's military boss for that first survey of the harbor was Captain George Gordon Meade, who just two years later would lead the Army of the Potomac to victory over Robert E. Lee at the Battle of Gettysburg.
NEMHC Portrait File

One Step Forward, Two Steps Back

The spectre of economic displacement once again created devastating social impacts on St. Louis County in 1873. Jay Cooke's plans to utilize Duluth as the terminus of his proposed Northern Pacific Railway across the northern border of the United States came to nothing when the Credit Mobilier scandal in Europe, the withdrawal of silver backing the nation's currency, and the debt load of most U.S. railroad bonds conspired to bring about the Panic of 1873.

The shaky financial situation of Cooke's Northern Pacific bonds received much of the popular blame for the onset of the Panic, and Duluth came in for its fair share of ridicule. One modern historian castigated Cooke's "patriotic cant about progress and the taming of the continent, the lavishly illustrated and fantastical brochures depicting Duluth as the Paris of the prairies, the public forums featuring local celebrities, [and] the traveling exhibits of the unexplored region's flora and fauna" as prime causes of the Panic.[12]

Whatever the cause, the results of the Panic were sharp, deep, and long-lasting. The U.S. economy was in the doldrums for the remainder of the 1870s, and unemployment stayed above 20 percent until 1880. Railroads were set back a decade; half of all U.S. rail bonds were worthless or in default by the time the United States celebrated its centennial in 1876.[13]

In Duluth and St. Louis County, the population once again took a backward leap, although not to the extent that the region had experienced during the Panic of 1857 sixteen years before. Still, in 1873 and 1874, the Lake Superior and Mississippi Railroad and the passenger packet ships that stopped at the Twin Ports carried far more people on the outbound portions of their journeys than on the inbound segments.

But St. Louis County recovered rapidly from the effects of the 1873 Panic, primarily because the port and county seat at the southeastern end of the county were quickly becoming one of the great grain ports of North America. From 1871 to 1880, the port at the western end of Lake Superior became the destination for much of the grain grown in Minnesota and the Dakota Territory to the west.

"A Great, Safe and Commodious Harbor"

Local citizens had helped dig a ship canal in the spring of 1871 to create access to the harbor, which could only be entered by the natural entry on the east side of nearby Superior.[14] The residents of the Wisconsin port city tried unsuccessfully to secure an injunction to have the Duluth Ship Canal halted, and when that proved to be too late, Superior embarked upon a decade-long attempt to close down the Duluth entry.

An 1872 storm that devastated Duluth's first lakeshore pier and wooden grain elevator showed city fathers the wisdom of having a harbor protected by the long sweep of Minnesota Point.[15] Judge Josiah Davis Ensign, who arrived at the Head of the Lakes in 1869 to practice law, wrote a history of the Duluth harbor nearly thirty years later, in 1898. Judge Ensign instinctively understood the role that the Duluth harbor would play in the growth and development of St. Louis County.

"Every resident of Duluth can give the particular reason that induced him to come here," Ensign wrote in 1898. "But the one primary reason that applied to all of the earliest settlers was the expectation that a great, safe and commodious harbor would be insured by the conditions that existed here; and the later comers were induced to come only by the fact that earlier comers had, with the conditions existing, realized their expectations by overcoming all obstacles and opposition, and has succeeded in making at Duluth the largest, most easily accessible and safest harbor on the great chain of lakes."[16]

The county's recovery from the Panic of 1873 was driven almost entirely by maritime commerce. By 1879, the Twin Ports of Duluth-Superior were handling nearly three hundred thousand tons of cargo, more than three times the tonnage handled in 1871. An even five hundred vessels called on the ports in 1879, and one-third of the cargo hauled down the Lakes—most to the growing terminal elevator complex in Buffalo, New York—was golden grain from the Minnesota prairies to the west and south of St. Louis County. In 1879, more than fifty thousand tons of coal flowed upbound to Duluth-Superior, fuel for the factories and mills that were spreading out from the Head of the Lakes.[17]

The completion of the Northern Pacific Railway to the Pacific Northwest in the late 1870s validated Jay Cooke's dream of linking the Great Lakes and the Pacific Ocean. The railway opened the vast wheat lands of the Red River Valley and Dakota Territory to cheap intermodal transportation through Duluth and the Great Lakes. A spate of grain terminal elevator construction on the east side of Rice's Point from 1878 to 1885 created the beginnings of Duluth's "elevator row" and fueled a virtual explosion of grain shipments through Duluth. The arrival of James J. Hill's Great Northern Railway in Superior in 1888 created an elevator-building boom on the Wisconsin side of the harbor from the mid-1880s to 1892.[18]

The result of all the building was the creation by the late 1880s of one of North America's fastest-growing grain shipping ports. Grain shipments went from eighty thousand tons in 1881 to just over 1 million tons in 1891. The average total tons passing through the Twin Ports in the 1880s was 1.4 million tons a year, ten times the average total tons handled by the ports in the 1870s. By the late 1880s, the Twin Ports were handling nearly 2.5 million tons of cargo each year, with upbound coal accounting for half the total tonnage each year from 1887 to 1889.

By the end of the decade, more than three hundred thousand tons of lumber were shipping out of the Twin Ports each year.[19] Much of the lumber was being cut on the North Shore and in the St. Louis River Valley upriver from Duluth. As timber cruisers tramped the flint hills above Lake Superior, they began to take renewed interest in the iron outcrops that had been widely known since the end of the Civil War.

This statue of Jay Cooke overlooks Superior Street east of downtown Duluth. Cooke, one of the most successful bankers in post–Civil War America, saw the potential of the Head of the Lakes. His vision for the Twin Ports as a rail and maritime hub was wiped out by the Panic of 1873, but was made a reality by other investors in the 1880s. Cooke's legacy is commemorated by the names of streets in Duluth's Lakeside neighborhood—Jay, Pitt, Cooke, Gladstone, Robinson, and McCulloch—which were named after the banker, his brother, and East Coast and British associates. NEMHC S2386 B8f24

George Calvin Stone came to Duluth as the president of Jay Cooke's bank. A "69er" who arrived in Duluth in the wave of settlement in 1869, Stone would spend the remainder of his life in Duluth. In 1880, the Duluth banker was instrumental in interesting other Philadelphia investors in the potential of St. Louis County's iron ore deposits. NEMHC Portrait File

George B. Sargent (1818–1875) served as surveyor general of the Minnesota, Wisconsin, and Iowa District before coming to live in Duluth in 1869. As the local representative of Philadelphia banker Jay Cooke, Sargent established the first banking house in Duluth and worked tirelessly to bring business and industry to the Head of the Lakes. NEMHC S3129 f3

In the spring of 1871, the citizens of Duluth took matters in their own hands and began the excavation of a ship channel between the Duluth-Superior harbor and Lake Superior just east of what is today downtown Duluth. Residents of Superior went to a federal court to secure an injunction against the new ship channel, arguing that it would cause the natural entry of the St. Louis River opposite Superior to silt up. The dredge Ishpeming *had done the lion's share of digging the ship channel. When Duluth citizens became aware that the injunction was on the way from Fort Leavenworth, Kansas, they turned out en masse with shovels, picks and dynamite to help the* Ishpeming *complete the last few yards. When the injunction arrived on Monday morning, traffic was steaming through the still unfinished ship channel.* Newton, NEMHC S2381 B1f3

Elevator A on the lakeshore at Fourth Avenue East was built in 1869 to handle the flood of grain flowing into Duluth aboard the freight cars of the Lake Superior and Mississippi Railroad. The massive wooden elevator was buffeted by the storms blowing in off Lake Superior. It served the growing grain transshipment business at the Head of the Lakes until Elevator Row was built along the east side of Rice's Point in the late 1870s and early 1880s. Gaylord, NEMHC S2395, v. 3, no. 12

Many of the residents of the North Shore who elected to stay in the wake of the Panic of 1873 scratched a living by commercial fishing. The western end of Lake Superior boasted huge stocks of lake trout, whitefish, and herring, and during the 1870s and 1880s commercial fishermen set out every morning from protected harbors along the Lake's North and South Shores. The catch was salted in barrels and shipped down the Lakes to grace the tables of Chicago, Cleveland, and Detroit restaurants and clubs. Gaylord, NEMHC S2395, v. 2, no. 137

German immigrant Frederick Weyerhaeuser made his first logging fortune in Wisconsin. Weyerhaeuser began investing in sawbucks in Cloquet and adjacent Itasca County in 1883, and by 1895 was in control of much of the timber in the St. Louis River Valley. Weyerhaeuser moved to St. Paul in 1891 and purchased a mansion on Summit Avenue next door to James J. Hill, the railroad magnate whose Great Northern spanned the continent from the Twin Cities to Seattle. The two captains of Gilded Age industry became best of friends in their later years. Ralph W. Hidy, Frank Ernest, and Allan Nevins, Timber and Men: The Weyerhaeuser Story, p. 177

Superior Street in 1870, seen here from Second Avenue West, was a sea of mud during most of the spring and framed for most of its short route by board sidewalks. One 69er later recalled that "Superior Street was a continuous succession of hills and gullies, connected its entire length by a four-foot plank sidewalk, with the planks laid endwise, bridging the ravines and tunneling the hills. To walk it was hazardous in the daytime, almost sure death after dark." General George Sargent, Jay Cooke's agent at the Head of the Lakes, was the driving force behind the 1870 construction of the Clark House, shown at left with the large flag flying atop the cupola, then the only hotel in the city. NEMHC S2386 B15f2

By 1885, Duluth reported a population of more than eighteen thousand people, up from fewer than thirty-five hundred people just five years before. The city's rapid recovery from the effects of the Panic of 1873 was primarily due to Duluth's commanding position astride the trade routes of the interior of North America. Grain from Minnesota flowed into the terminal elevators lining the harbor and was shipped down the Great Lakes to the mill complex at Buffalo, New York. By the mid-1880s, steam-driven wooden vessels were capable of hauling three thousand tons of grain down the Lakes from Duluth in a single trip, twenty times the tonnage capacity of the ships that plied the Lakes just two decades before. NEMHC S2386 B15f5

Alfred R. Merritt, shown here in about 1896 with his second wife, Jane, came to Duluth as a child and remained at the Head of the Lakes through the Panic of 1857 and the Panic of 1873. With his brothers Leonidas and Cassius, and various nephews and cousins, Alfred R. Merritt explored much of central St. Louis County in the 1880s and 1890s. NEMHC Ely Studio; S3120 B18, no. 209

James Jerome Hill was a Canadian who came to St. Paul in the mid-1850s. During the next four decades, Hill turned the bankrupt St. Paul and Pacific Railroad into the transcontinental Great Northern Railway. The Great Northern extended its lines to Lake Superior in 1885, although Hill bypassed Duluth for Superior. In the 1890s, however, Hill would play a key role in the development of St. Louis County's almost limitless iron ore resources. Albro Martin, James J. Hill and the Opening of the Northwest

The population explosion in Duluth in the 1880s necessitated a boom in school construction. In 1887, when the newly built Adams School opened just above the Point of Rocks west of downtown, the local school system was educating more than sixteen hundred students in eight separate schools. The city's growing Roman Catholic population also educated students in several parochial schools. NEMHC S2386 B8f7

The construction of the Duluth High School in 1887—seen here with the student body two years later—was a landmark event in the county's educational history. The new high school at First Avenue East and Third Street was called a "model of its kind." Later the Liberty School, the three-story brick facility was replaced by the Central High School several blocks west in 1892. Robert E. Denfeld, a graduate of Amherst College, came to Duluth as the city's school superintendent in 1885. Denfeld remained in Duluth the next thirty years and put his indelible stamp on the Zenith City's school system. NEMHC S2386 B8f13a

Duluth's growth through the mid-1880s flowed from the central and east hillsides east along the harbor and Minnesota Point. Still a village in 1887, Duluth retired many of the bonds that had defaulted during the Panic of 1873 and elected its first mayor, John B. Sutphin. The city also embarked on a series of municipal improvements, creating a city park system and paving Superior Street with treated cedar blocks. NEMHC S2386 B15f5

J. C. "Buzz" Ryan, the much loved amateur historian of northern Minnesota's lumber industry in the late nineteenth and early twentieth centuries, noted that "the Duluth mills built high tramways along the waterfront and piled their lumber so that it could be loaded directly on to the boats. The railroads built docks along the waterfront at which trainloads of lumber coming from inland mills could be loaded directly on to the boats." The growth of lumber shipments through the Twin Ports was phenomenal. In 1885, a little over 1,000 tons of lumber moved across the docks into lumber packets for Lower Great Lakes ports. By 1888, almost 400,000 tons of lumber went down the Lakes from Duluth, much of it to Tonawanda, New York, near Buffalo. Tonnages dropped off sharply from 1889 to 1894, reached 440,000 tons the next year, and doubled to 940,000 tons by 1899. NEMHC S2386 B26f1

The economic engine that drove Duluth's expansion during the 1880s was Elevator Row, located along the east side of Rice's Point, seen here in a wide-angle photograph from the hillside in the mid-1880s. More than three thousand workers were employed to build the eleven elevators on Elevator Row between 1878 and 1885, and several area sawmills were dedicated exclusively to turning out the dimension lumber used to construct the huge structures. When the Northern Pacific Railway completed its Elevator G in 1885, Elevator Row had a combined storage capacity of some 9 million bushels of wheat. Gaylord, NEMHC S2395, v. 2, no. 133

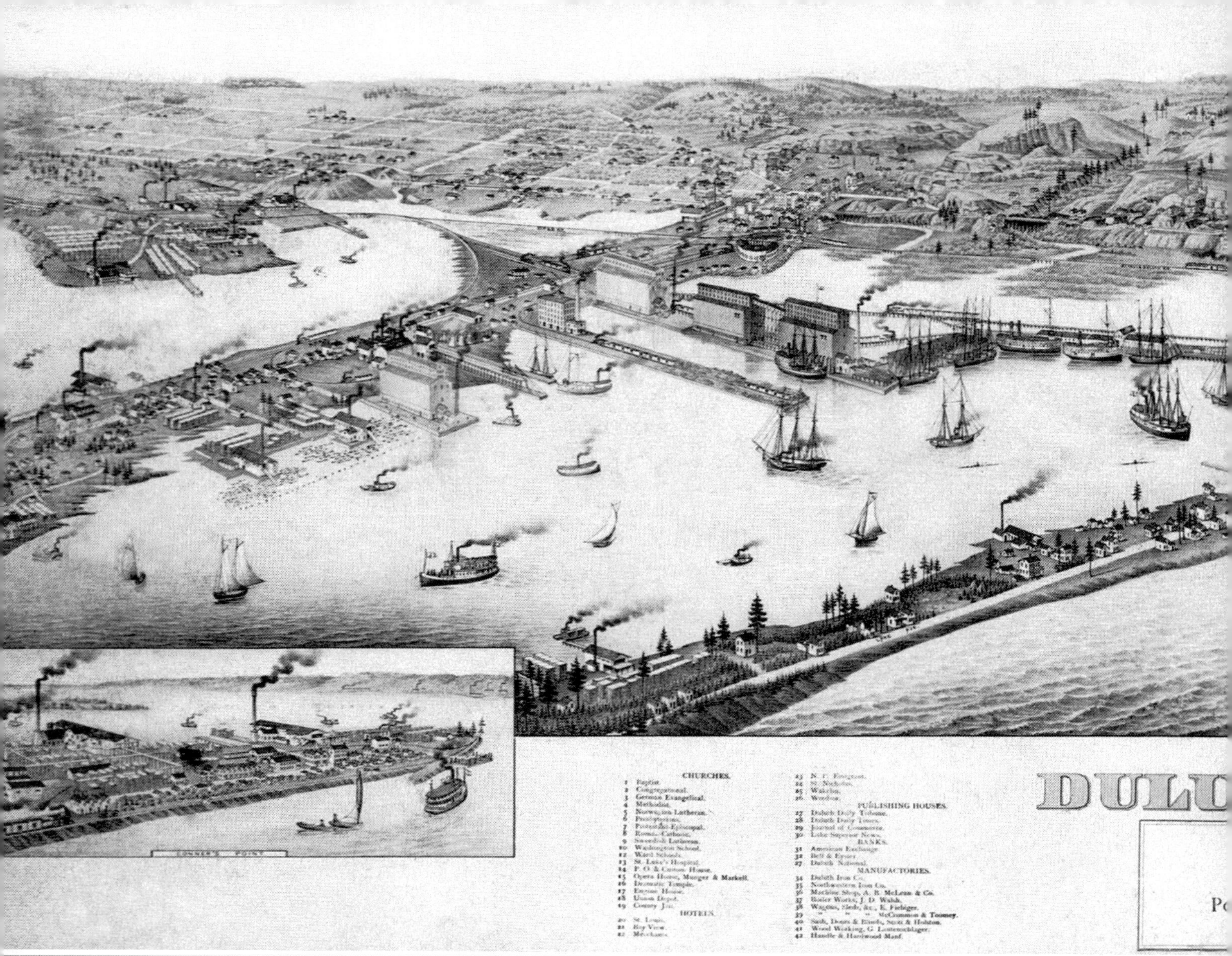

CONNER'S POINT
CHURCHES.
1 Baptist.
2 Congregational.
3 German Evangelical.
4 Methodist.
5 Norwegian Lutheran.
6 Presbyterian.
7 Protestant Episcopal.
8 Roman Catholic.
9 Swedish Lutheran.
10 Washington School.
12 Ward Schools.
13 St. Luke's Hospital.
14 P. O. & Custom House.
15 Opera House, Munger & Markell.
16 Dramatic Temple.
17 Engine House.
18 Union Depot.
19 County Jail.
HOTELS.
20 St. Louis.
21 Bay View.
22 Merchants.
23 N. P. Emigrant.
24 St. Nicholas.
25 Wakelin.
26 Windsor.
PUBLISHING HOUSES.
27 Duluth Daily Tribune.
28 Duluth Daily Times.
29 Journal of Commerce.
30 Lake Superior News.
BANKS.
31 American Exchange.
32 Bell & Eyster.
27 Duluth National.
MANUFACTORIES.
34 Duluth Iron Co.
35 Northwestern Iron Co.
36 Machine Shop, A. R. McLean & Co.
37 Boiler Works, J. D. Walsh.
38 Wagons, Sleds, &c., E. Fiebiger.
39 " " " McCrimmon & Toomey.
40 Sash, Doors & Blinds, Scott & Holston.
41 Wood Working, G. Lautenschlager.
42 Handle & Hardwood Manf.
DULU

The Duluth harbor was a beehive of activity in 1884. What had been a shallow harbor bisected by a dike ten years before, the result of a settlement between Duluth and Superior concerning the dredging of the Duluth Ship Channel in 1871, was a thriving hub of maritime commerce in 1884. Almost five hundred vessels called on the Twin Ports in 1884, delivering nearly 375,000 tons of coal and loading more than 325,000 tons of Minnesota grain. The total of 886,000 tons was double the slightly more than 400,000 tons handled through the harbor just four years before; by 1888, four years in the future, the Twin Ports would be moving almost 2.5 million tons of cargo. NEMHC S2386 B48 "0"

The Duluth Minnesotian *advertised itself as the "Official Paper of St. Louis County." Thomas Foster produced the weekly newspaper every Saturday from April 24, 1869, to September 4, 1875, including this issue on July 18, 1874.* NEMHC rm160 5/B/7

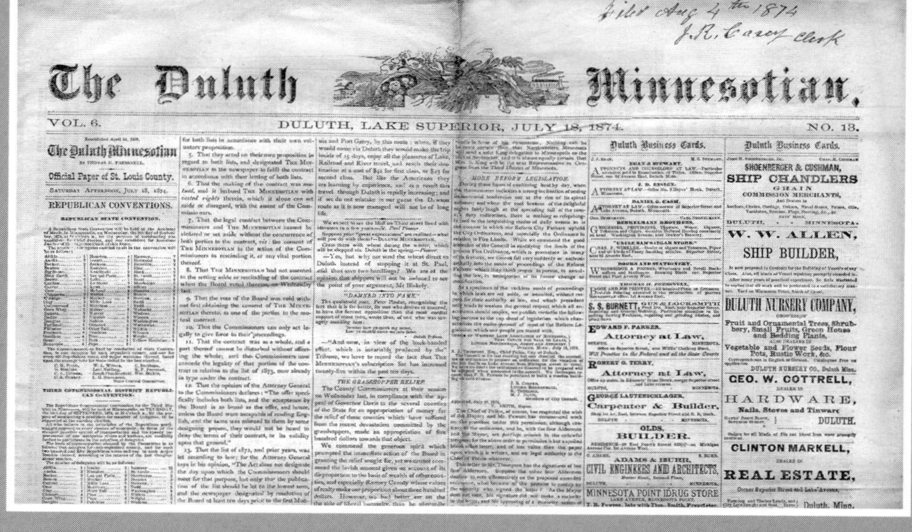

Filed Aug 4th 1874
J. R. Carey clerk

The Duluth Minnesotian.

VOL. 6. DULUTH, LAKE SUPERIOR, JULY 18, 1874. NO. 13.

The Duluth Minnesotian

Official Paper of St. Louis County.

SATURDAY AFTERNOON, JULY 18, 1874.

REPUBLICAN CONVENTIONS.

REPUBLICAN STATE CONVENTION.

THE GRASSHOPPER RELIEF.

MORE REFORM LEGISLATION.

"DAMNED INTO FAME."

Duluth Business Cards.

EGAN & STEWART.

J. D. ENSIGN.

DANIEL G. CASH.

HERKELMANN BROTHERS.

"UNCLE SAM'S CIGAR STORE."

BOOKS AND STATIONERY.

S. S. BURNETT.

EDWARD F. PARKER.

Attorney at Law,

ROBERT G. TERRY.

Attorney at Law,

GEORGE LAUTENSCHLAGER,

Carpenter & Builder,

OLDS,

BUILDER.

ADAMS & BURR,

CIVIL ENGINEERS AND ARCHITECTS.

MINNESOTA POINT DRUG STORE

Duluth Business Cards.

SHOENBERGER & CUSHMAN,

SHIP CHANDLERS

GRAIN

COMMISSION MERCHANTS,

DULUTH, - - MINNESOTA.

W. W. ALLEN,

SHIP BUILDER,

DULUTH NURSERY COMPANY,

Fruit and Ornamental Trees, Shrubbery, Small Fruits, Green House and Bedding Plants,

Vegetable and Flower Seeds, Flour Pots, Rustic Work, &c.

GEO. W. COTTRELL,

HARDWARE,

Nails, Stoves and Tinware

DULUTH.

CLINTON MARKELL,

REAL ESTATE,

Duluth, Minn.

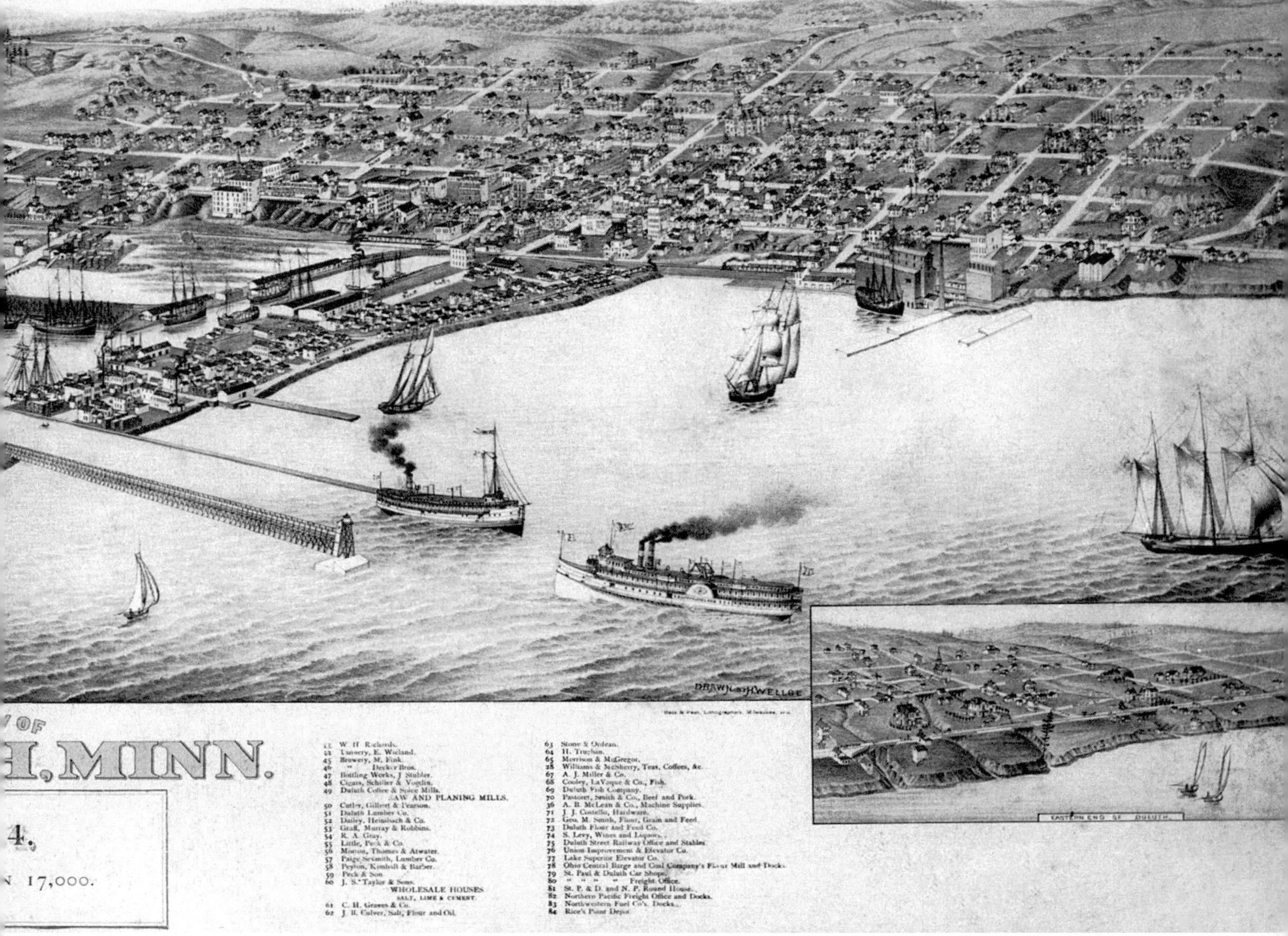

V OF

H, MINN.

4,

N 17,000.

DRAWN BY H. WELLGE

Beck & Pauli, Lithographers, Milwaukee, Wis.

44 W. H. Richards.
44 Tannery, E. Wieland.
45 Brewery, M. Fink.
46 " Decker Bros.
47 Bottling Works, J. Stabler.
48 Cigars, Schiller & Vogtlin.
49 Duluth Coffee & Spice Mills.

SAW AND PLANING MILLS.

50 Cutler, Gilbert & Pearson.
51 Duluth Lumber Co.
52 Dailey, Heimbach & Co.
53 Graff, Murray & Robbins.
54 R. A. Gray.
55 Little, Peck & Co.
56 Moron, Thomas & Atwater.
57 Paige, Sexsmith, Lumber Co.
58 Peyton, Kimball & Barber.
59 Peck & Son.
60 J. S. Taylor & Sons.

WHOLESALE HOUSES

SALT, LIME & CEMENT.

61 C. H. Graves & Co.
62 J. B. Culver, Salt, Flour and Oil.
63 Stone & Ordean.
64 H. Trughin.
65 Morrison & McGregor.
28 Williams & McSherry, Teas, Coffees, &c.
67 A. J. Miller & Co.
68 Cooley, LaVaque & Co., Fish.
69 Duluth Fish Company.
70 Pastoret, Smith & Co., Beef and Pork.
36 A. B. McLean & Co., Machine Supplies.
71 J. J. Costello, Hardware.
72 Geo. M. Smith, Flour, Grain and Feed.
73 Duluth Flour and Feed Co.
74 S. Levy, Wines and Liquors.
75 Duluth Street Railway Office and Stables.
76 Union Improvement & Elevator Co.
77 Lake Superior Elevator Co.
78 Ohio Central Barge and Coal Company's Flour Mill and Docks.
79 St. Paul & Duluth Car Shops.
80 " " " " Freight Office.
81 St. P. & D. and N. P. Round House.
82 Northern Pacific Freight Office and Docks.
83 Northwestern Fuel Co's. Docks.
84 Rice's Point Depot.

EASTERN END OF DULUTH.

By 1890, Duluth was on its way to becoming the world's busiest lumber port. As the timber industry depleted the white pine forests of the Lower Peninsula of Michigan and northern Wisconsin, the loggers moved west into Minnesota and St. Louis County. By 1885, most of the St. Croix Valley south of Duluth had been logged over, and timber cruisers were assessing merchantable stands of pine on the North Shore of Lake Superior and along the St. Louis River Valley west of Duluth. Most of the wood cut along the North Shore and in St. Louis County was hauled from landings by logging railroads to one of the dozen mills in Duluth. By 1900, logging railroads were hauling more than 3 million board feet of logs a day into the Twin Ports. NEMHC S2386 B26f1

The Minnesota Point Lighthouse shone a welcome beacon across the frequently turbulent waters of western Lake Superior. NEMHC S2386 B30f19

The propellor Northern Light *was a regular caller on Duluth in the early 1870s. She is seen here docked at the outer harbor with the brig* Commerce *in 1872.* NEMHC S2395 v. 3, no. 3

Top: The brig Commerce *with her sails set, departing Duluth for the Soo, about 1872. The age of sail on Lake Superior lasted well into the 1880s.* Gaylord, NEMHC S2395, v. 3, no. 5

The propellor Chicora was built in England in 1864 and saw duty on the Upper Lakes from 1869 to 1877. She was shifted to the Lower Lakes in the late 1870s and ran goods and passengers between Niagara and Toronto until the 1920s. Gaylord, NEMHC S2395, v. 3, no. 6

Mr. John M. Butler,

requests the pleasure of your company to inspect
the St. Louis River Slate Brick Company
of Thomson, Minnesota,

and to meet

Mr. Jay Cooke of Philadelphia.

Special train will leave Union Depot, Duluth, at V *o'clock, P. M.*
June 22, 1893; returning will reach Duluth about Six P. M.

R. S. V. P.

An 1893 invitation to meet the "great man" himself. Cooke had been long retired to his Lake Erie mansion by that time. NEMHC S4572, no. E214

Duluth's Point of Rocks, seen here in 1887, was a local landmark from the beginning. NEMHC Duluth Illustrated 1887

The Lake Superior and Mississippi Railroad freight depot in Duluth at the lakefront and Fourth Avenue East. The arrival of the railroad in the early 1870s ensured that Minnesota farmers had another outlet for their grain. NEMHC S2395, v. 2, no. 95

Crews building the Northern Pacific Railway into Duluth. Jay Cooke's dream of connecting Duluth to the Pacific Ocean in the early 1870s was not realized until a decade later, but when it did happen, Duluth became one of the nation's great grain ports. NEMHC S2395, v. 2, no. 91

Iron Empire

chapter three

Until the late 1880s and early 1890s, the population of St. Louis County basically...

Until the late 1880s and early 1890s, the population of St. Louis County basically was confined to a ten-mile radius of the Lake Superior port city of Duluth.

The reason for the population concentration was simple. The rugged country north and west of Lake Superior had yet to be penetrated by the railroads. The simplest way of getting about was little changed since the days of the Anishinabe and the *voyageurs*. Waterborne transportation, whether by canoe on the region's rivers or by lake steamer on Lake Superior, was the quickest and cheapest method of travel, at least north and west across the county or up the North Shore of Lake Superior.

Because of the growing rail network in the 1880s, county residents could easily get to the Twin Cities and points south, or to Fargo-Moorhead and points west. And the rise of passenger packet ships on the Great Lakes in the 1880s meant that county residents could travel to Buffalo, Cleveland, Detroit, and points in between safely, swiftly, and in comfort during the nine-month navigation season.

But traveling anywhere in the interior of St. Louis County meant walking for the most part, or traveling by horse-drawn cart or dog sled during the winter. There was a trail north to the Bois Fort Reservation and pioneer settlement on Lake Vermilion, and the old canoe route across the northern boundary of the county still existed in 1890. The Canadian Pacific Railway completed its trackage from Winnipeg and Fort Garry to Fort William and Port Arthur on Lake Superior in 1882, opening up land north of the international boundary for settlement in the mid-1880s.[1]

By 1885 most of St. Louis County and adjacent Itasca and Koochiching Counties to the west and north were encircled by rail networks. In the late 1880s, timber

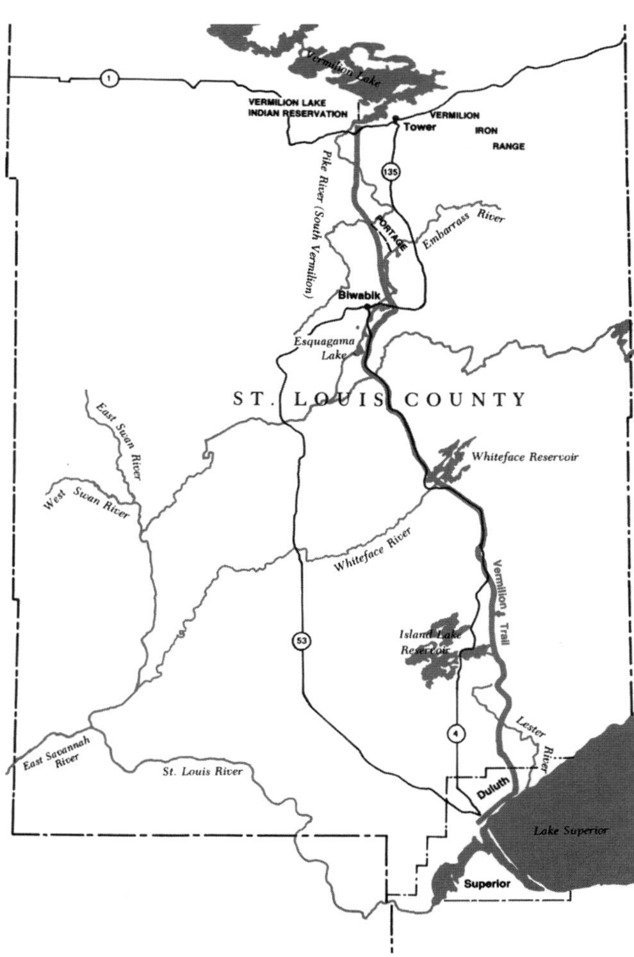

In the years before steel rails crossed St. Louis County, the only way to get to Lake Vermilion in the north end of the county was via the Vermilion Trail, a crudely cut path north from the east end of Duluth to the outlet of Pike River on the lake. Island Reservoir and the Whiteface Reservoir, which the Trail skirts and crosses today, didn't exist in the mid-nineteenth century when gold-seekers and timber cruisers explored the mineral potential of northern St. Louis County. Map, Minnesota History, v. 44, no. 2, p. 48

companies began to penetrate the arrowhead region of northeastern Minnesota in search of white pine. Timber cruisers ranged across the rocky hills north and west of Lake Superior in search of stands of virgin white pine. Along the way, they recorded their observations of the landscape, the flora, and fauna. Most had one eye cocked for gold and other precious metals, but the relatively common mineral they found would make the county's fortune for the next century-and-a-quarter.

George Stuntz, who had been tramping the Lake Superior country for more than twenty years, had first collected iron ore samples from south of Lake Vermilion in 1865, during the brief gold rush. Stuntz had packed a bag with more than sixty pounds of samples and hauled it on his back to Duluth, but no one was interested.[2]

The time just wasn't right. With the Civil War winding down, America was making the painful conversion back to a peacetime economy. The nation's infant steel industry was still located in places like Pennsylvania's Lehigh Valley, and hot iron was largely made in beehive furnaces. The discovery of massive deposits of iron ore on what would become known as the Marquette Range in Michigan's Upper Peninsula gave North America's steel industry a reserve of raw material that experts predicted might take decades to exhaust.

Stuntz kept grabbing anybody who would listen to tell them about "that big mountain of iron" he suspected existed on the shores of Lake Vermilion.[3] One who lent a receptive ear in the 1870s was George Calvin Stone, who had come to Duluth in 1869 as Jay Cooke's local banker. Stone, who was fifty years old at the time of the 1873 Panic that short-circuited his boss's grandiose plans for the Head of the Lakes, believed Stuntz's tales of an iron mountain in the interior north and west of the Big Lake.

By 1875, the nation was beginning to recover from the effects of the Panic two years before. Paradoxically, the booming railroad industry that had precipitated the Panic was driving much of the recovery. Building railroads across the sparsely settled prairies of the West and Northwest opened up by the homestead legislation of 1864 required thousands of tons of steel rails.[4]

Englishman Henry Bessemer's process for producing steel made its way across the Atlantic Ocean in 1870s. The much more efficient Bessemer process rapidly replaced the beehive charcoal furnaces that had been used to make steel since the early 1800s.

Scotsman Andrew Carnegie broke ground for the first modern steelmaking plant in the United States at Braddock, Pennsylvania, in 1873. By the time the Edgar Thomson Works began pouring steel in 1875, America was on its way to becoming the world's pre-eminent steel producer. During the

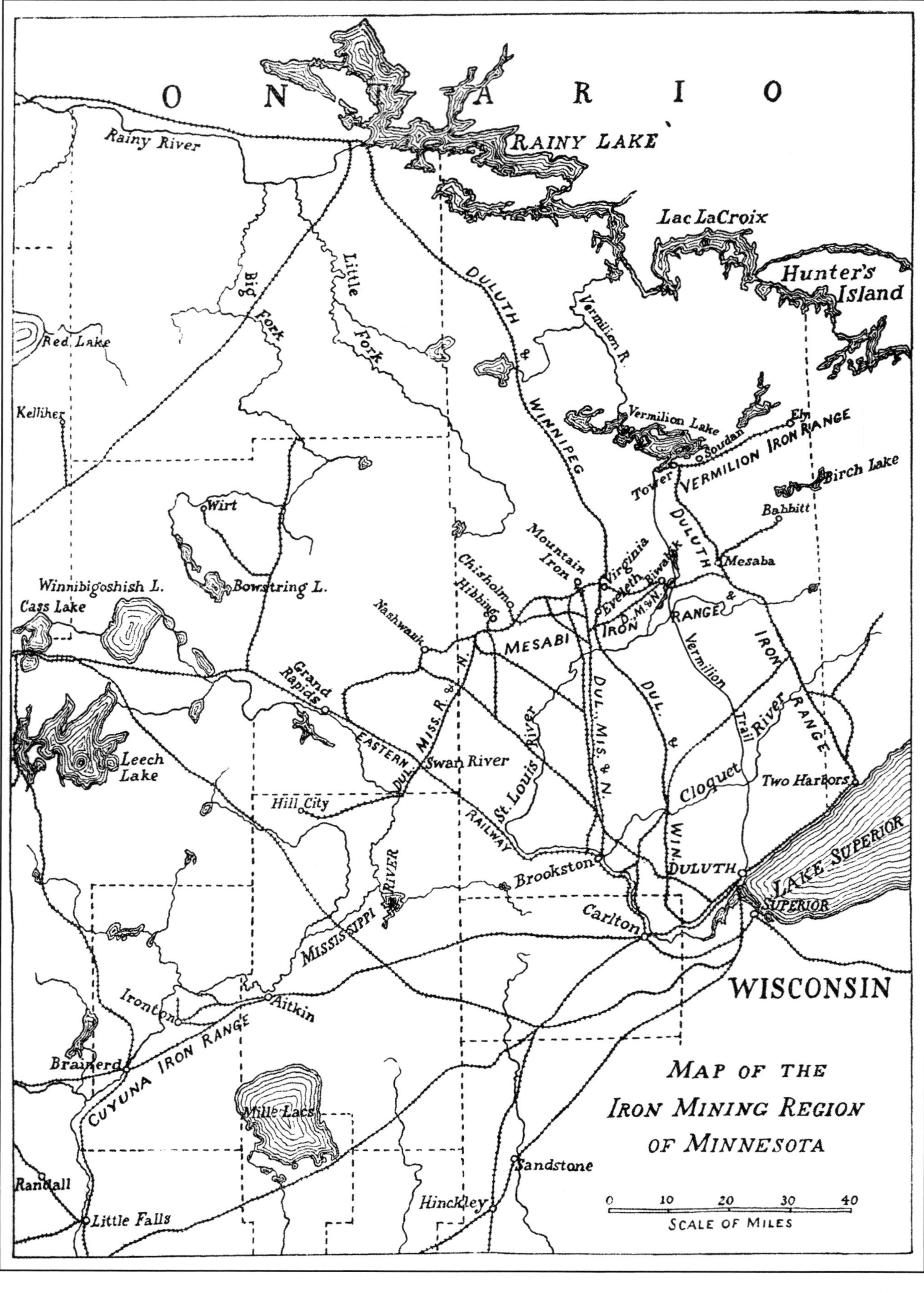
ONTARIO
Rainy River
RAINY LAKE
Lac LaCroix
Hunter's Island
Big Fork
Little Fork
DULUTH & WINNIPEG
Vermilion R.
Red Lake
Kelliher
Vermilion Lake
Soudan
Ely
VERMILION IRON RANGE
Tower
Birch Lake
Babbitt
Wirt
Mountain Iron
Virginia
Eveleth
Biwabik
DULUTH
Mesaba
Chisholm
Hibbing
Winnibigoshish L.
Bowstring L.
Cass Lake
Nashwauk
MESABI
IRON
RANGE
D. M. & N.
RANGE &
IRON
Vermilion
Grand Rapids
DUL. MISS. R. & N.
DUL. MIS. & N.
DUL. & WIN.
Trail
River
RANGE
Leech Lake
EASTERN
Swan River
St. Louis River
Cloquet
Two Harbors
Hill City
RAILWAY
Brookston
DULUTH
LAKE SUPERIOR
SUPERIOR
RIVER
MISSISSIPPI
Carlton
WISCONSIN
Ironton
Aitkin
CUYUNA IRON RANGE
Brainerd
MAP OF THE IRON MINING REGION OF MINNESOTA
Mille Lacs
Sandstone
Randall
Hinckley
Little Falls
0 10 20 30 40
SCALE OF MILES

Opposite page: By the early 1880s, residents of St. Louis County could easily reach just about anywhere in Minnesota and the Northwest south of a line from Duluth to Fargo-Moorhead. Rail lines crisscrossed the southern half of the state, making convenient connections between Duluth and the Twin Cities, as well as between Duluth and the Red River Valley. A resident of Duluth, however, who wanted to reach the settlements on Rainy Lake would take the train to Fargo, where they could head north by Red River cart or on a Red River steamboat to Rat Portage. At Rat Portage, later to be named Kenora, they would change to another steamboat for a trip down Lake of the Woods and up the Rainy River to what is today International Falls. The trip could easily take several weeks. Folwell, v. 4, opp. p. 8

next quarter-century, Carnegie would make the Monongahela River Valley just outside Pittsburgh into North America's most productive steel district.[5]

The expansion of the steel industry in the years following the Civil War created a demand for iron ore that hadn't been foreseen when George Stuntz hauled his sack of samples back to Duluth from Lake Vermilion in 1865.

Charlemagne Tower Makes an Investment

While the United States was recovering from the effects of the Panic in 1875, Duluth and Superior were still reeling from the impact. Cooke was gone from the scene, splitting his time between appearing in court on the East Coast and hiding from creditors at his mansion on the shores of Lake Erie. But a second, wealthy Eastern capitalist picked up the mantle of Cooke's dreams and made possible the development of a vast swath of northern St. Louis County.

Charlemagne Tower was an 1830 Harvard College graduate who had made a fortune in Pennsylvania anthracite during the mid-nineteenth century. An attorney, he had known Jay Cooke for years through his investments in railroad bonds sold by the Philadelphian's banking house. Tower had invested heavily in Northern Pacific Railway stock and was on the company's Board of Directors when the Panic of 1873 washed away most of the value of his investment.

Tower was named a trustee of the Northern Pacific as the railroad sought to work its way out from bankruptcy proceedings, and to his credit, the Pottsville, Pennsylvania investor never lost his faith in the development potential of the railroad. Tower was well aware of the rumors that said northern Minnesota was home to a mountain of iron ore, and Duluth banker George Stone had visited him in early 1875 to urge Tower to look more closely at the iron potential of the Lake Superior region. Tower had asked an old friend, Albert Chester, a New York geology professor, to investigate the country near Lake Vermilion. Chester and Richard Henry Lee, Tower's son-in-law, spent the summer of 1875 surveying the Lake Superior country west of Duluth.

Accompanied by George Stuntz, the geological survey party investigated the south shore of Lake Vermilion and then tramped across the Mesabi Hills south and west. The party spent more than a month surveying the area between Partridge Lake and Birch Lake. They found iron ore—plenty of it—but it was a leaner variety than the deposits that soon would be discovered even farther south and west.[6]

Chester wasn't the first to examine the iron ore geology of the country between Birch Lake and Lake Vermilion. In the early 1870s, just after the gold rush, Christian Wieland and several of his brothers surveyed south of Lake Vermilion and brought back iron ore samples to their lumber camp at the mouth

A native of Utica, New York, Charlemagne Tower completed his legal training in 1836 and eight years later helped a family friend purchase coal properties in the anthracite region of eastern Pennsylvania. The friend took the young attorney in as a partner, and Tower moved to Pottsville, Pennsylvania in 1851 to oversee his coal investments. A millionaire many times over, Tower sold his coal properties in 1871 and invested the proceeds in railroad and mining bonds. When George Calvin Stone approached Tower in 1875, the Pennsylvanian was already familiar with the possibility of commercial iron ore deposits in northern Minnesota. A director of Jay Cooke's Northern Pacific Railway Company, Tower was one of the co-trustees charged with guiding the railroad out of bankruptcy following the Panic of 1873. NEMHC Portrait File

of Beaver River on Lake Superior north of Duluth. The five Wieland brothers operated a sawmill at Beaver Bay and traded with the mining communities across the Lake at Ontonagon, Copper Harbor, and Marquette.[7]

The Wieland brothers interested a syndicate of Ontonagon and Duluth investors in the iron ore potential of Lake Vermilion. The syndicate incorporated as the Mesaba Iron Company in 1874 with former Minnesota Governor Alexander Ramsey as president, and even received a land grant from the Minnesota Legislature to build a railroad north from Duluth to Lake Vermilion.[8]

The Panic of 1873 and Professor Chester's initial report in the summer of 1875 that the iron ore north of Birch Lake on what would become the eastern end of the Mesabi Range was too lean for Bessemer furnaces halted talk of a mountain of iron north of Duluth. But Chester sent Stuntz back to

Charlemagne Tower Jr.'s management team included two brothers-in-law, Isaac P. Beck and Richard H. Lee. Beck served as treasurer of the Duluth and Iron Range Railroad from 1882 to 1884, and Lee was the railroad's chief engineer during the same period. Lee had degrees in civil and mining engineering and had accompanied Professor Albert Chester on his first visit to Lake Vermilion in 1875. Lee established offices in Agate Bay—later Two Harbors—and supervised the construction crews from John S. Wolf and Company of Ottumwa, Iowa, the low bidder on the railroad construction project. NEMHC Portrait File

Crews built the station at Tower in 1885, the first depot erected on the Duluth and Iron Range. More than fourteen hundred people lived in Tower at the time, and more were flocking in to Soudan, the residential community adjacent to Tower. Franklin King, Mesabi Road, p. 31

Lake Vermilion in late July 1875, and the samples that the Duluthian brought back assayed at 67 percent iron ore, more than satisfactory for reduction in Bessemer furnaces.[9]

Tower asked Chester to return to Lake Vermilion in 1880. This time, the New Yorker reported that the deposit was huge and the samples assayed as much as 70 percent iron. Best of all, the iron was low in phosphorous, which meant it was perfect for feeding to Bessemer furnaces.

The Vermilion Range

Charlemagne Tower used Chester's 1880 report to justify a major investment in Minnesota iron lands. Tower instructed Stone to begin buying land just south of Lake Vermilion, and the Pennsylvania investor quickly unveiled ambitious plans to develop iron mines in northern St. Louis County. In 1882, Tower incorporated the Minnesota Iron Company for $10 million, an immense sum at the time. The next year, Tower sent Franklin Prince, a mining engineer, to the Vermilion Range to begin building the infrastructure for a mining venture.[10]

Tower acquired a railroad—the Duluth and Iron Range—from the Wielands and the Ontonagon syndicate. He interested brothers Samuel and George Ely from Marquette in investing in the new iron ore development. Meanwhile, Prince built a sawmill at the mine site in St. Louis County and began laying out the town site of Tower. By early 1884, a full work crew was at the site, and Charlemagne Tower and his shareholders had already spent $3.5 million on the new mine and railroad.[11]

Even more important than development of the mine site was the work on the Duluth and Iron Range Railway, which would link the iron ore deposits on the south end of Lake Vermilion with the all-important lake shipping industry on the North Shore of Lake Superior. Originally, the plan had been to build the railroad from Tower to Duluth, primarily because the state's land grant for building the railroad stipulated a southern terminus at Duluth.

But the difficulty of laying track across the rough country south of Tower necessitated a change. Tower and his investors eventually decided to take a much shorter route to Lake

Tower in 1885 was the biggest community in St. Louis County outside of Duluth. In this photo taken looking east along the business district, wooden sidewalks elevated above the streets kept shoppers from wading through mud and horse manure. Unlike many western mining camps of the time, Tower from the start was a family community; already by 1885 females made up more than 30 percent of the population. Part of the reason for the family orientation involved the Cornish miners brought to the community by Captain Elisha Morcom from Michigan's Menominee Range to work the mines. The Cornishmen were the world's most accomplished miners and typically came to a mining frontier with their wives and children. NEMHC S2386 B20f4

Superior, following an old Anishinabe trail from Tower to Agate Bay—which would later be renamed Two Harbors.[12]

In 1883 and 1884, some six hundred workers swarmed across the route from Tower to Agate Bay, clearing a right-of-way and laying track. Before the last spike was driven, Tower and his investors would spend $2 million for railroad construction. Meanwhile, Prince and his work crew at Tower were busy digging a pit and building a town. By 1884, nearly fourteen hundred people lived in the new town of Tower. Additional people were moving into the neighboring community of Soudan, named after the British colony south of Egypt in a tongue-in-cheek comparison to the brutally cold winters of northern St. Louis County.[13]

Finally, on July 31, 1884, the first locomotive and ore cars chugged up the trackage from Agate Bay to Tower. The ten ore cars were loaded with the first shipment of Minnesota hematite ore, and the train headed back down the hill to Lake Superior, sixty-eight miles away. It arrived at Agate Bay at midnight, and dumped its load of ore into the pockets of a new twelve-hundred-foot ore dock the next morning. Before August 1, 1884, was over, the steamer *Hecla* and schooner *Ironton* had headed downbound to Sault Ste. Marie with the first load of Minnesota iron ore consigned to the American steel industry.[14]

Ironically, the cost of developing an iron range from raw wilderness came close to breaking Charlemagne Tower. The Pottsville attorney who had made millions of dollars in Pennsylvania anthracite spent millions of dollars on Minnesota iron. In 1887, a syndicate of Chicago investors, including Cyrus McCormack and the executives of two of the Windy City's biggest steel mills, began buying land north of Tower. The syndicate purchased more than twenty-five thousand acres near what would become the town of Ely and soon announced plans to build a new railroad to Lake Superior—if they couldn't buy the Duluth and Iron Range.

Tower, by then in his late seventies and yet to recoup his massive investment in Minnesota iron mining, began negotiations. In the fall of 1887, just three years after shipping the first load of ore to Agate Bay, Tower sold his interest in the Minnesota Iron Company and the railroad to the Chicago syndicate for $8.5 million. He nearly doubled his original investment with the sale and kept a $500,000 interest in the syndicate's Minnesota Mining and Railroad venture.[15]

Opening the Mesabi

Charlemagne Tower wasn't the first investor to be temporarily overwhelmed by the financial magnitude of developing iron ore resources in Minnesota. Duluth's Merritt family wrestled an iron range from the wilderness and lost it to Eastern investors, all in the space of little more than five years.

Lewis Merritt and his family had arrived at the Head of the Lakes before the Panic of 1857. Like Stuntz, Merritt and his sons made their living in the woods. Four of the sons and two nephews were timber cruisers and knew the rugged country north and west of the Twin Ports like the backs of their hands.

Lewis J. Merritt and his brothers, Leonidas, Alfred, Cassius, and Napoleon, spent much of the 1880s exploring the same ground that Stuntz and Chester identified for Charlemagne Tower. In the spring of 1889, the Merritts set out southwest from Lake Vermilion and Tower-Soudan. They carried shovels, picks, and black powder, and spent the early spring digging test pits along their route. What they found was a soft, red ore that assayed at 60 percent iron.[16]

Before news of the discovery became too widespread, the Merritts managed to assemble 140 land leases in the area of their iron discovery. The family suspected it controlled an ore body far richer than those already developed at Tower-Soudan and even farther north at the frontier community of Ely. But the Merritts lacked the capital to sink the mine pits, build a railroad, and erect ore docks at a Lake port. Their task in 1890 was to find a Charlemagne Tower willing to risk the huge amounts of investment capital needed to develop the potential of their iron ore discovery.

Henry Clay Frick, a Pennsylvania coal millionaire allied with Andrew Carnegie, passed on the opportunity to invest in the Merritt's find. The family found investors in St. Paul and incorporated the Duluth, Missabe and Northern Railway Company to build a rail line from their mine at Mountain Iron to Stony Brook Junction, the terminus of the Duluth and Winnipeg Railway just outside the Twin Ports.[17]

The first school opened in Tower in 1884. By the time this picture was taken in 1889, the community was levying $8,500 for schools, including $4,500 in salaries for teachers and the principal. Teachers at the time were paid $60 a month, while the principal for School District No. 9 received $90 a month.
NEMHC S2386 B20f4

The Merritts made an important Pittsburgh connection in the summer of 1892 when they convinced Henry W. Oliver of the richness of the ore from the Mesabi Range. Oliver, a Monongahela Valley steel mill owner and Carnegie associate, was in the Twin Cities to help nominate Benjamin Harrison as the Republican Party's standard-bearer in the 1892 presidential election. Intrigued by the reports from St. Louis County, he ordered his rail car hauled north to Duluth, where he met with the Merritts and visited their mine. Impressed by the workings and the richness of the ore, Oliver committed to taking delivery of four hundred thousand tons of ore within two years' time.[18]

The Oliver contract opened the door to more Pittsburgh investment in the Merritts' venture. By October 1892, the first load of ore from the Mesabi Range came down the completed

The ore stockpile at Tower began building during the spring and had usually been exhausted late in the year. In the early years of mining, ore was hauled to the docks at Two Harbors in twenty- to twenty-five-ton cars, at about three hundred tons per trainload. An increase in the size of locomotives and rail cars allowed operators to ship far more tonnage per trainload, and by 1892, more than 1 million tons of ore went down to Lake Superior from the Vermilion Range. Newton, NEMHC S2381 B1f1

The Chandler Mine established Ely as an integral part of the Vermilion Range, primarily because of its rich Bessemer ore. "It is a very desirable ore," one mining engineer wrote in the late 1880s, "and finds a ready purchaser." Newton, NEMHC S2381 B1f1

rail line from Mountain Iron. The ore went out of the Duluth and Winnipeg's small dock at Allouez on the east side of Superior, and the Merritts soon started looking for suitable land in Duluth to build an ore dock.

First, however, they were approached by the Minnesota Iron Company, the company in which Charlemagne Tower still retained an interest, with an $8 million offer for the family's holdings on the Mesabi. The Merritts turned the offer down and began searching for financing to extend the Duluth, Missabe and Northern Railway from Stony Brook to a mammoth new ore dock at Fortieth Avenue West in Duluth.[19]

Wall Street financier Charles Wetmore loaned the Merritts the money for the rail extension and new ore dock, and by the spring of 1893, construction was well under way. But once again, the specter of economic dislocation played a key role in the development of St. Louis County. The collapse of the Philadelphia and Reading Railroad precipitated the Panic of 1893. By summer, the nation's economy was in shambles.

The Merritts were victims of bad timing. Like Jay Cooke twenty years before, they happened to be in the wrong place at the wrong time. In the summer of 1893, most of the contractors that they had hired to extend the railroad and build the ore docks had laid off their crews. Wetmore, the financier, was unable to come up with all of the money he had promised.

Even worse, one of the investors he had recruited for the loan to the Merritts was John D. Rockefeller, the Cleveland oil baron. To recoup his investment, Rockefeller and his associates approached the Merritts and offered to lend them another $100,000 of operating capital. Desperate for cash, the Merritts agreed and set up a joint holding company with Rockefeller to own and operate the mine, railroad, and ore dock.[20]

Before 1893 was over, Rockefeller had pumped $2 million into Lake Superior Consolidated Iron Mines, and in effect, controlled the holding company. The Cleveland industrialist and the Merritts had a major falling out, and the partnership soon wound up in court. After a series of publicized trials, charges, and countercharges, Rockefeller paid the Merritts $525,000, and the Duluth family recanted its charge that the Cleveland industrialist was guilty of fraud.[21]

Rockefeller himself soon sold his interest in the Mesabi Range mines discovered by the Merritt family, although he remained a power in the Great Lakes shipping industry for the next twenty years. Henry Oliver and his partner, Andrew Carnegie, gained control of some of the Mesabi's richest mines. When J. P. Morgan put together U.S. Steel Corporation in 1901—the world's first $1 billion corporation—the new conglomerate was built on Mesabi ore.

The development of Minnesota's first two iron ranges—the Vermilion and the Mesabi—accomplished two very important things for St. Louis County. The flow of iron ore, which totaled well over 5 million tons from the Mesabi alone by 1899, created an infrastructure of railroads, roads, and mining communities across the previously inaccessible central and northern sections of the county. St. Paul capitalist James J. Hill pushed a rail line to the Mesabi Range by the turn of the century.

Towns such as Virginia and Hibbing—named for Frank Hibbing, a German-born timber cruiser who sold mining interests on the Mesabi to Rockefeller in 1893—mushroomed across what residents would soon simply refer to as "the Range." Immigrants would flood into the Range towns during the first decade of the twentieth century and create the unique culture that characterized St. Louis County during the next one hundred years.

Secondly, iron ore made Duluth, the county seat, into a world-class port. Rockefeller and his successors built three huge ore docks in West Duluth between 1893 and 1900.[22] Nearby Superior also benefited, and iron ore transformed the neighboring cities into true "twin ports." James J. Hill's Great Northern Railway acquired control of the Duluth and Winnipeg ore docks in Allouez in 1899 and immediately built a second, much larger ore dock.[23]

In 1893, the Merritts shipped just over five hundred thousand tons of Mesabi ore down the Lakes through the Twin Ports. Seven years later, in the first year of the new century, nearly 6 million tons flowed down the Lakes from Duluth-Superior.[24] The county seat's population was fifty-three thousand in 1900, a gain of more than 300 percent in just fifteen years.

St. Louis County would enjoy equally strong growth during the first two decades of the twentieth century.

The Chandler Mine, the Pioneer Mine, and the other Vermilion Range mines around Tower-Soudan were all initially deep-shaft mines. The Chandler Mine, seen here in 1887, comprised four shafts, two to the north of the main ore deposit and two to the south. Mining was done on a caving system, which its operators claimed resulted in very low costs for underground mining. Newton, NEMHC S2381 B1f1

Duluthian Robert B. Whiteside, a fee owner of several of the mines in the Ely district, built the town's first hotel, the Pioneer, in 1887. At the time, Ely's business district also included a post office, grocery store, and a hardware and furniture store.
NEMHC S2386 B16f1

Long before sled dog racing became a sport in northern Minnesota, dog teams were one of the best ways to get from Ely to nearby communities during the winters of the early twentieth century.
NEMHC S2386 B20f4

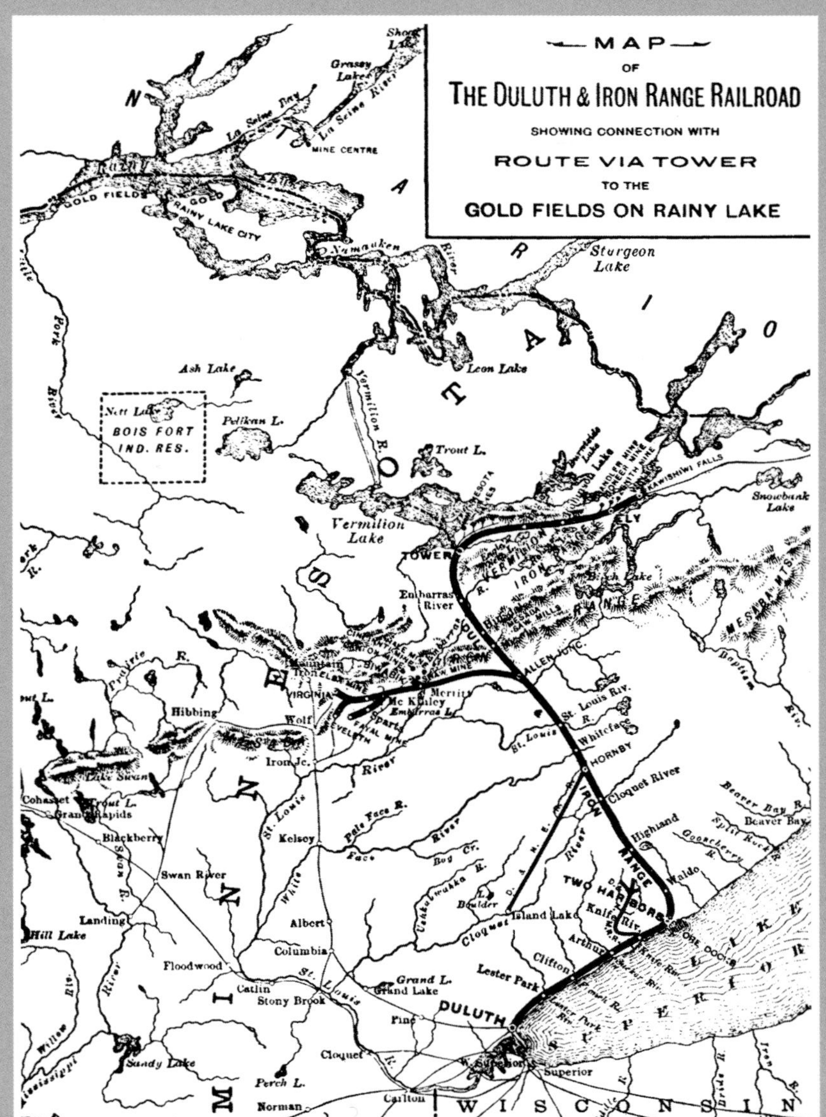

Left: The Duluth and Iron Range Railroad that Richard Henry Lee completed in 1885 opened up the northern third of St. Louis County to Duluth and the more settled south end of the county. In 1886, the railroad completed an extension from its ore docks at Two Harbors to Lester Park and Duluth. The railroad and its successors operated the North Shore spur for the next hundred-plus years. Map, Franklin King, Missabe Road, p. 40

Cassius Merritt was seven years Lon Merritt's junior. In his 20s, Cassius served as deputy St. Louis County auditor, treasurer and register of deeds. He later operated a grocery store in Superior before embarking on a career as one of the most accomplished timber cruisers and estimators in northern Minnesota. In 1887, while cruising timber for a rail right-of-way near what is today Mountain Iron, he stumbled and fell upon a chunk of nearly pure iron ore "about as big as his two fists." Cassius Merritt carefully stowed the piece of ore in his pack and brought it back to Duluth for assay. NEMHC S3120 B28, no. 1959, no. 2099

Leonidas Merritt was one of eight sons of Lewis H. Merritt who lived to adulthood. Lon, as he was known in the family, moved to Duluth at the age of twelve and spent the remainder of his life as a timber cruiser, explorer, and prospector. Lon and his brothers Alfred, Napoleon, and Cassius are credited with the initial development of the Mesabi Range from their first mine near Mountain Iron. NEMHC S3120, no. 594

Hulett C. Merritt, the son of Lewis J. Merritt, left northern Minnesota for southern California in the late 1890s. He got involved in the electric utility industry, built a seven-thousand-acre ranch near Pasadena and died a millionaire in 1956. NEMHC, Merritt, Vertical File

Lewis H. Merritt, the son of Alfred, was another of the second-generation members of the family who worked so hard to make the Mesabi Range a reality. NEMHC S3120 B19, no. 197

Alexander McDougall was a Scot who grew up in Canada. After a twenty-year career sailing the Great Lakes with the Anchor Line, McDougall came ashore at Duluth and began building steel freighters at a shipyard in Howard's Pocket on the Superior side of the harbor. McDougall called his unique design a "whaleback," for its low profile that gave it a stable ride in the rough seas of the Great Lakes. Detractors called it a "pigboat" for its flat snout, but McDougall's whaleback design was immediately popular for shipping iron ore. When Mesabi ore began coming down the hill into the Twin Ports, the first whalebacks had just come down the ways. Between 1888 and 1898, more than forty whaleback steamers and consort barges were launched at the Head of the Lakes. The vast majority was immediately pressed into service hauling the ever-increasing flood of iron ore downbound to the nation's steel mills on the Lower Great Lakes. NEMHC Portrait File

Henry W. Oliver's visit to Minnesota in the summer of 1892 was the key to initial development of the Mesabi Range. A boyhood friend of Thomas Carnegie, Andrew Carnegie's brother, Oliver had been a driving force in the Bessemer furnace segment of the Monongahela Valley steel industry since just after the Civil War. By the turn of the twentieth century, he would control much of the iron ore production on the Mesabi Range. Iron Pioneer, NEMHC Portrait File

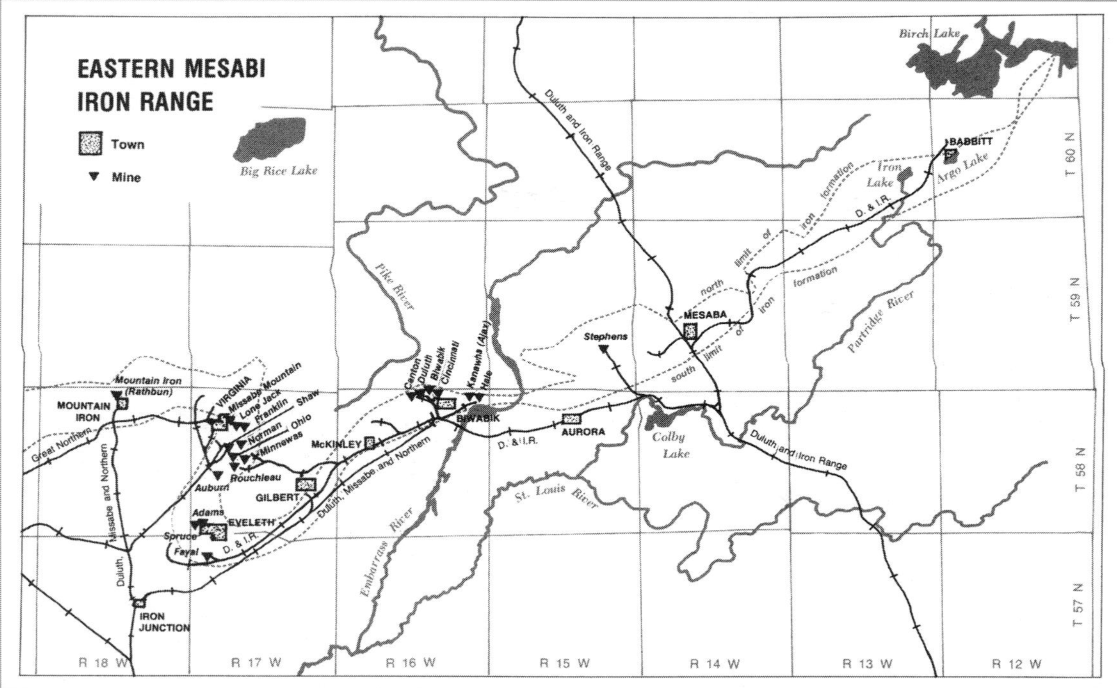

Top: The eastern Mesabi yielded the Range's first iron ore in the fall of 1892. The Merritts built their Duluth, Missabe and Northern Railway from Mountain Iron south to Stony Brook Junction. Later in the 1890s, the consortium who grabbed control of the Mesabi extended the DM&N to Virginia and from Iron Junction through Eveleth to Biwabik. David Walker, Iron Frontier, Map, p. 92

The first load of iron ore from the Mesabi Range passed through Duluth to the Duluth and Winnipeg Railway dock at Allouez in late October 1892. Much like topping out a building, the evergreen tree atop the ore car signified the completion of the construction phase of the mining development. NEMHC S2386 B29f4

Top: An ore dock is a marvel of nineteenth century industrial engineering. Built almost entirely of wood, the ore dock combined a high trestle with interior bins. A train full of ore cars was driven across the trestle and the bottom-dump cars emptied into the bins, or pockets. A Great Lakes vessel was then positioned alongside the dock face. Its holds were lined up with chutes, which funneled the ore in the gravity-feed pockets into the holds. The Merritt family's decision to build a new ore dock at Fortieth Avenue West in Duluth brought about the collapse of the family's mining venture during the Panic of 1893. NEMHC S2386 B41f11

The interior of a whaleback hold. The chief drawback to the whaleback was its small size. McDougall could not build a whaleback with a beam, or width, of more than forty-five feet without compromising cargo space by building interior supports and struts. By early in the twentieth century, the more traditional lake vessels had a beam of seenty-five feet and could routinely carry ten thousand tons of iron ore in a single load. NEMHC S2396 B16, v. 40, no. 24

In the early 1900s, a whaleback steamer ties up to the Mesabi Docks in West Duluth and is surrounded by a half-dozen or more traditional lake vessels. Just ten years into the Mesabi Range era, the docks in West Duluth already were handling in excess of 10 million tons of iron a year. NEMHC S3742 B7f23

The opening of the Mesabi Range created an unprecedented need for labor. Within ten years of the first iron ore shipment from Mountain Iron, thousands of men were hired to build railroads, mines, and ore docks; strip the open pits for mining; load rail cars; and maintain millions of dollars of equipment. Here, a crew of men uses long poles to loosen the ore in rail cars above the Duluth, Mesabi and Northern docks in West Duluth. During the spring and fall, moisture in the ore would sometimes cause the ore to freeze in the cars, and manual labor was needed to break the ore loose so it would fall into the dock pockets. NEMHC S2386 B39f19

Crews repair a battery-operated "mule" on one of the lower levels of Ely's Pioneer Mine, around 1890. Most of the Vermilion Range mines in Ely and neighboring Tower-Soudan were underground, rather than open pit. NEMHC S2386 B29 f9a

John D. Rockefeller as a young man. After creating a virtual monopoly in the nation's infant oil business, the Cleveland capitalist quickly moved to control the shipment of iron ore on the Great Lakes. NEMHC Portrait File

Andrew Carnegie was a dour Scotsman who created the modern American steel industry. Carnegie retired from the business in 1900 when J. P. Morgan bought him out and created U.S. Steel, the nation's first billion-dollar corporation. Carnegie spent the rest of his life giving his money away to build libraries across America. NEMHC Portrait File

Ely's Pioneer Mine, Shaft A, looking northwest, about 1890. The headframe in the foreground contained powerful hoisting machinery that lowered miners to the ore bearing lode and brought iron ore back to the surface. NEMHC S2386 B29 f9a

Whalebacks under construction at Alexander McDougall's shipyard at Howard's Pocket in Superior in the 1890s. Whalebacks were never elegant in the water, but Great Lakes sailors swore by them. NEMHC S2396, v. 40, no. 13

1906 Interlude

By 1906, the trickle of iron ore from the Mesabi and Vermilion Ranges had become a veritable cascade, transforming every segment of life in St. Louis County. Areas that had been wilderness less than twenty years before were thriving mining communities, linked to the wider world by a network of rail lines and electric street railways. Settlers from every nation in Europe were flocking into the county, attracted by the prospect of jobs that paid far more than agricultural work back in Italy, Finland, or Serbia.

Duluth, with a population approaching sixty thousand people, was fast becoming the "Pittsburgh of the North," with crews from U.S. Steel beginning to clear a huge plot of land on the western edge of the city for a mammoth new steel mill. Steam whistles at the city's ore docks, lumber mills, factories, and grain elevators blew twenty-four hours a day. In the harbor, dredges contracted to the U.S. Army Corps of Engineers were in the midst of a ten-year harbor improvement project that would make the Twin Ports one of the most productive harbors on the Great Lakes.

The growth of towns on the Mesabi and Vermilion Ranges was little short of phenomenal. Tower, Soudan, Ely, and Winton dated from the middle to late 1880s. McKinley, Biwabik, Virginia, Hibbing, Eveleth, Sparta, and Mountain Iron on the Mesabi Range were chartered between 1892 and 1895. By 1906, all were thriving communities that continued to add jobs and families.

Hibbing, founded in 1895 on a one-hundred-acre plot at the west end of the Mesabi's then known ore deposit, grew from one thousand people in its first year to six thoiusand people the next. By 1906, town boosters claimed a population of almost fifteen thousand.[1] Driving the surge in population was the growth of mining on the Mesabi and the insatiable demand for iron ore at the nation's steel mills. Within walking distance of Hibbing were sixteen iron ore mines, including the Hull-Rust and Mahoning, which would come to define the town named for German timber cruiser Frank Hibbing.[2]

The towns of the Mesabi Range were spaced so close together that even by 1906, it seemed that the area was far more urban than it really was.

Thomas F. Cole was perhaps the most influential person in St. Louis County in 1906. Cole, who grew up on the Michigan Iron Ranges, was president of the Oliver Iron Mining Company in 1906. At the time, he and two other Duluthians, Guilford G. Hartley and Chester Congdon, were involved in opening up the Canisteo District on the western Mesabi Range. The town of Coleraine in Itasca County would be named in Cole's honor. NEMHC Portrait File

Chisholm, with a population of more than four thousand, was just eight miles east of Hibbing, twenty minutes away by train and less than an hour in the newfangled automobiles, or horseless carriages, that were beginning to make their presence known in St. Louis County. Buhl and Kinney were just seven miles east of Chisholm, and Mountain Iron was another eight miles farther on. Virginia was three miles east of Mountain Iron—practically next door—and Eveleth was seven miles south of Virginia. McKinley and Biwabik on the eastern end of the Mesabi were more distant—thirteen to fifteen miles east of Virginia.

Virginia was the commercial center of the Mesabi. Burned out twice, in 1893 and 1900, the city rebuilt each time, and in 1906, Virginia's commercial district was erected entirely of stone, brick, and concrete. The city boasted a modern electric light plant and had an estimated population of twelve thousand people.[3]

Another county resident who lent his name to a Range community was Frank Hibbing. A German immigrant and timber cruiser, Hibbing sold a half-interest in several Mesabi Range mining claims for $250,000 and retired to Duluth, where he died in 1897. NEMHC Portrait File

The rapid development of the Mesabi Range had been the driving force in the county's phenomenal growth. In just over fifteen years, the county population had more than doubled—from just under forty-five thousand people in 1890 to an estimated one hundred thousand or more in 1906.[4] What's more, in 1890, the population had been concentrated in Duluth and in a narrow swath around Tower-Soudan and Ely. By 1906, the population was spread out across the Mesabi Range, the Vermilion Range, and in Duluth. Meanwhile, settlers were filling in the spaces in between along the rail lines that connected the three population centers.

Praying in Seven Languages

The compact size of the Mesabi Range—it was only forty-four miles from Hibbing to Biwabik—meant that the area already had a sense of regional identity and community. In 1906, everything was hustle and bustle. Mine pits were being dug, houses were being thrown up, businesses were opening, all across the Range.

Perhaps the biggest news that summer was the completion of the rail spur between Hibbing and Grand Rapids to the southwest that opened up what was then called the Canisteo District. In reality the western extension of the Mesabi Range, the Canisteo was a beehive of activity, with steam shovels taking big gulps of the rich, red earth, and mining locations springing up at Bovey, Nashwauk, Keewatin, and Coleraine.[5]

An increasing number of the residents flooding into the older communities on the eastern and central Mesabi and the infant mining locations in nearby Itasca County were immigrants. These new immigrants would give the Iron Range the polyglot culture it has enjoyed for more than a century. Unlike earlier waves of immigrants who populated Minnesota and Wisconsin from Germany, Scandinavia, and Ireland, these new immigrants were from provinces in southern and western Europe that most Americans could barely pronounce, let alone locate on a map.

A huge crowd gathers for the laying of the cornerstone at the Eveleth City Hall on June 16, 1906. The $20,000 building housed the city's police and fire departments on the ground floor and administrative offices on the second floor. When Eveleth later built a new fire hall, city hall was completely dedicated to city departments. NEMHC S2386 B16f3

They came from the Balkans, fleeing poverty and blood feuds in Serbia, Croatia, Slovenia, Bosnia, and Montenegro. They came from the sulfur mines of southern Italy and the sun-baked villages of Sicily, areas still recovering from the *risorgimento* and civil wars of the nineteenth century. They came from the Czar's duchy of Finland, young men escaping from forced service in the Russian armies being decimated by the Japanese on the other side of the world.

Many of the county's new residents were recruited by labor agents in Trieste or Turku or Stockholm or Palermo. For passage money to the United States, they agreed to work two years or five years for something called the Oliver Iron Mining Company in a place called Minnesota. Many others came to St. Louis County via other mining districts or communities.

Perhaps a majority of the county's Finnish residents came to Duluth and the iron ranges from Hancock in nearby Houghton County, Michigan, where they had gained mining experience working for the Quincy Mining Company or the mighty Calumet and Hecla Consolidated Copper Company. Others came from port communities such as Ashtabula and Sandusky on Lake Erie, or from the iron mining ranges on Michigan's Upper Peninsula.

The immigrants settled in places like Chisholm. The Range community had a population of about forty-three hundred people in 1906. Two-thirds of the residents—twenty-seven hundred people—were foreign-born, the highest percentage of any community in St. Louis County.[6] The immigrants weren't always welcomed by the earlier residents of the county. Those from the Balkans were called Austrians, to denote the fact that in the old country, they were subjects of the Austro-Hungarian emperor. To many of the residents of the Range, Finns were Finlanders, and Italians from whatever region they hailed were commonly called dagoes.

A poem that celebrates the richness of Mesabi Range diversity recalls miners who "prayed in seven languages."[7] It was said of Rangers that grandma could speak, read, and write Serbian, and curse and pray in English. Her daughter was typically bilingual, speaking both Serbian and English fluently. Her granddaughter could speak, read, and write English, and curse and pray in Serbian.

Rangers brought their customs with them from the Old World. Finns gave the range *kallamoiiaka,* their spicy fish stew. Serbs were proud of their sweet cakes, the *potica* that so many other Range nationalities appropriated for their own tables. And every ethnic group learned how to make the Cousin Jacks' *pasties,* the meat and vegetable pie that fit over the thermos of hot tea in a miner's steel lunch pail.

The Cousin Jacks, however, didn't stay on the Range long enough to influence the culture the way they had in nearby Michigan. People with a heritage of deep-shaft mining dating back to Roman Cornwall, most had come to the Vermilion with Captain Elisha Morcom twenty years before. The mines surrounding Tower-Soudan and Ely originally were deep-shaft operations, where the Cornish miners felt comfortable working. But the Mesabi Range, from the beginning, was open-pit mining, a form of mineral extraction the Cousin Jacks disdained. "Goddam ditch digging," one Cornish miner is supposed to have uttered on his way out of town for more congenial diggings in the Sierra Nevada.

Captain Martin Trewhella was one of the few Cousin Jacks who successfully made the switch from deep-shaft to open-pit mining. Born in Cornwall in 1861, Trewhella came to the United States as a young man and found work in the coal mines of Pennsylvania. In 1881, Trewhella migrated to Ishpeming, Michigan, to work in the mines of the Marquette Range. He helped sink shafts in Montana's Butte Copper District and was on the Mesabi Range when the twentieth century opened. In 1906, Trewhella was the mine captain for International Harvester's Agnew Mine. Van Brunt, v. 3, p. 1116

Many of the immigrants also brought with them political philosophies that were far more radical than the two-party system in the United States thought appropriate at the turn of the twentieth century. Matti Halberg arrived on the Mesabi Range around 1900 already indoctrinated with socialist political principles in his native Finland. In 1906 or shortly thereafter, Halberg joined the Industrial Workers of the World (IWW), formed that summer in Chicago. In years to come, Matti Halberg, who then farmed near Cherry, became a charter member of the Communist Party USA and enrolled his son, Arvo Kuusta Halberg, in the Young Communist League.[8] Years later, the younger Halberg, who anglicized his name to Gus Hall, recalled with a laugh that he and his father had been the Communists in the family, often arguing with his brothers, who remained loyal to the IWW and its philosophy of syndicalism.[9]

The radicalism of many of the immigrants would translate into state and local politics and the often-contentious labor-management relations in St. Louis County during the twentieth century. In 1906, some of those Finns and Austrians already were quietly meeting with labor organizers from the Denver-based Western Federation of Miners (WFM) to discuss complaints about working conditions in the area mines. The Federation would fail in an organizing drive on the Mesabi the following year. In 1908, the WFM would be catapulted into world headlines when William Dudley "Big Bill" Haywood, a founder of both the WFM and the IWW, and two other executives of the Federation were arrested and charged with the murder of former Idaho Governor Frank Steunenberg.

If the radical politics of the iron ranges gave St. Louis County a wild and wooly reputation in 1906, Duluth, the county seat, was rapidly becoming one of the more cosmopolitan cities in the Upper Midwest.

Zenith City of the Unsalted Seas

While the rapid growth of the Iron Range communities in the late 1890s and early 1900s was little short of phenomenal, the county's population was still centered in Duluth. With an estimated population of close to seventy thousand people in 1906, Duluth was the second-largest metropolitan area in Minnesota and among the premier cities of the Northwest at the time. The city's population comprised more than half of St. Louis County's population at the time.

The city's industrial base was predicated on iron ore, timber and transportation. In 1906, more than 19 million tons of iron ore moved through the Twin Ports in rail cars, destined for the iron ore docks of Duluth and Superior. That was more than three times the iron ore that flowed across the Twin Ports' docks in 1900.[10]

The rapid expansion of iron ore shipments from Duluth and Superior necessitated almost constant improvement of the two cities' ore docks. In 1906, the Great Northern Railway was finalizing plans for a major rebuilding the next summer of its Dock No. 1 in Allouez, first erected in 1892.[11] In the summer of 1906, the Duluth hillside resonated to the sounds of sawing and hammering as an army of workers built docks and wharves for sand, gravel, package, freight and dozens of other commodities.

George L. Brozich was typical of the upward mobility of the new immigrants to St. Louis County. Born in Serbia in 1878, Brozich, his siblings, and mother followed his father to the Michigan Copper Country in 1888. In 1890, the family moved to the Vermilion Range, where Brozich and his brother completed a high school education. Brozich finished a two-year course in commercial law and banking at St. John's College in Collegeville and then spent five years in the retail business in Colorado and Illinois. In 1906, Brozich had returned to the Vermilion Range where he worked as assistant cashier at the Exchange Bank in Ely. Van Brunt, v. 3, p. 977

Several thousand residents of the Twin Ports made their living in the communities' coal trades. Between 1906 and 1909, coal companies serving the region built coal docks aggregating more than 2.5 million tons of annual capacity, mostly on the Superior side of the harbor.[12] In the 1906 navigation season, more than 5 million tons of coal were off-loaded from Great Lakes steamers at the coal docks of the Twin Ports.[13] Coal fueled the early twentieth-century industrial revolution, and an increasing number of county residents opted for the convenience of a coal-fired heating stove in their living quarters.

Grain was another staple of Twin Ports' maritime commerce in 1906. Nearly 1.7 million tons of hard red spring wheat from Minnesota and the Dakotas went down the Lakes in 1906 to the massive flour-

The link that connected the Mesabi Iron Range to Duluth was the ore dock complex at Fortieth Avenue West. By 1906, the Duluth Missabe and Northern, which operated the Duluth ore docks, and the Duluth and Iron Range, which delivered its Vermilion Range ore to docks at Two Harbors, were both subsidiaries of U.S. Steel Corporation. NEMHC S2386 B29f8

milling complex clustered around the port of Buffalo, New York.[14] Just six years before, Frank Peavey of Minneapolis had demonstrated the feasibility of storing grain in concrete elevators at a site on Rice's Point in Duluth.[15] The concrete elevator, which had the capacity to store 3.35 million bushels of wheat, joined other grain terminal elevators on the east side of Rice's Point, with a capacity of some 13 million additional bushels.

Timber was the final driver of the local maritime economy in 1906. More than seven hundred thousand tons of sawtimber went down the Great Lakes in 1906, most of it cut from St. Louis County and from logging sites located along both the North and South Shores of Lake Superior.[16] Duluth and Superior rocked twenty-four hours a day with the sound of buzz and gang saws from the area sawmills. All that timber was destined for residential housing and office buildings in other growing Great Lakes port cities such as Milwaukee, Chicago, Detroit, Cleveland, and Buffalo.

The Duluth Aerial Bridge was already the signature structure of the Twin Ports when this photo was taken in 1906. Modeled after a similar bridge in Rouen, France, the Aerial Bridge included a man car which could hold a streetcar or several automobiles for the one minute, forty-second crossing of the Duluth Ship Canal. Gallagher, NEMHC S2386 B1f8

The Zenith Furnace Company was one of the iron-based industries that popped up in Duluth in the 1890s and early 1900s. In 1906, the firm occupied an eighty-acre site in West Duluth on St. Louis Bay and used coking coal delivered to the Twin Ports by boat to make about 350,000 tons of pig iron from Mesabi Range ore. McKenzie, NEMHC S2422 B1f20

In 1906, the U.S. Army Corps of Engineers was winding down a harbor improvement project that had been ongoing since the late 1890s. The project included reconstructing the Duluth Ship Canal and the Superior Entry, dredging seventeen miles of channels in the harbor to a twenty-foot depth and creating a protected anchorage basin in the harbor.[17]

The fifty-six hundred vessels that called on the Twin Ports in 1906—even though an ever-increasing number were built of steel and powered by steam—were still at the mercy of Lake Superior's fierce storms. Duluth residents had watched horrified the previous November as the steel bulk freighter *Mataafa* broke up off the outer entry of the Ship Canal. Nine of the vessel's crew froze to death in plain sight of downtown Duluth.[18]

The iron economy of the Twin Ports created dozens of new manufacturing businesses that employed thousands of workers. What would become Clyde Iron was founded in 1899 to produce, among other things, stamp hammers for lumberjacks in the Great Lakes states.[19] Zenith Dredge, which incorporated in 1905, was in the process in 1906 of locating its dipper dredge and three dump scows at the foot of Thirteenth Avenue West.[20] Just two blocks away, young Peter Grignon was in the process of taking over his uncle Napoleon's interest in the Marine Iron and Ship Building Company.[21]

Trevanion W. Hugo was probably Duluth's most influential politician in the early part of the 1900s. Three times mayor of Duluth in the first twenty years of the twentieth century, Hugo defeated incumbent Mayor Henry Truelson by six votes in the 1900 election and by eight votes in his attempt to regain the office in 1902. The incumbent was derided as "Typhoid" Truelson for his delay during a series of epidemics in the late 1890s, and Hugo campaigned on a platform of providing the city with a world-class water treatment plant. Truelson, however, was later noted for his efforts to open up city government to the people during his term. NEMHC Portrait File

The prosperity of the Twin Ports in 1906 spawned big dreams for the communities' future. In 1906, the Great Northern Power Company was in the process of completing the Thomson Dam on the St. Louis River above Duluth. The dam, which would be the world's third largest when placed into commercial operation in 1907, fed its power to a mammoth brick substation at Fifteenth Avenue West in Duluth.[22] From there, the electricity would be distributed to grain elevators, ore docks, sawmills, factories, and the Duluth Edison Electric Company, which early in 1906 was organized to provide electricity to Duluth's residential and commercial customers.[23]

Duluth had all the conveniences of a modern city: a growing downtown overlooking the lake, paved streets and sidewalks thronged with traffic, electric streetcars, a world-class water treatment system, and pleasant neighborhoods for the city's residents. The city also was rapidly developing a reputation as a regional health care center for all of northern Minnesota and northwestern Wisconsin.

Duluth boasted two first-rate hospitals just several blocks apart. St. Luke's Hospital had been founded by the rector of St. Paul's Episcopal Church in 1881 and had moved to Second Avenue East and Fourth Street three years later. In 1900, the hospital moved to its present site at Ninth Avenue East and First Street. The hospital's trustees' erected a modern, four-story brick hospital in 1902.[24]

Albert M. Marshall made a fortune in the mining industry selling hardware to the iron mining companies of the Mesabi and Vermilion Ranges. An Ohioan, Marshall arrived in the Twin Ports from Michigan just before the onset of the 1893 Panic. He bought a half-interest in a Duluth wholesale hardware distribution firm. By the turn of the twentieth century, the Marshall-Wells Hardware firm was one of the three largest distribution firms in the United States, serving retail and wholesale customers from the Rocky Mountains to Wisconsin. NEMHC Portrait File

St. Mary's Hospital had gotten its start in the 1880s when Benedictine Sister Amata crisscrossed the forests and bogs of northern Minnesota, selling the "lumberjack ticket," a primitive health insurance policy that entitled the bearer to health care at the Sisters' hospital in West Duluth.[25] In 1898, the Benedictine Sisters moved the facility to its current site at Fifth Avenue East and Third Street and built a state-of-the-art two-hundred-bed hospital.[26]

St. Louis County's prosperity in 1906 was built on a foundation of iron. That foundation seemed all the more unshakeable the following year when U.S. Steel Corporation announced that it was going to spend nearly $6 million to build a steel mill and company town at Morgan Park on the far west end of Duluth.[27] The "steel trust," as many in St. Louis County called it, bowed to pressure in the Minnesota Legislature for onerous new tonnage taxes on iron ore shipped out of state unless the Pittsburgh conglomerate agreed to locate a manufacturing facility in Minnesota.[28] As it was, it took U.S. Steel more than eight years to build the new mill, but it began rolling steel in late 1915, just as the demand for steel skyrocketed because of World War I.

Civic boosters predicted a bright future for Duluth and St. Louis County. "The great tributary country, of which Duluth is the natural gateway," an editorial writer for the *Duluth News-Tribune* noted in 1910, "is settling up, increasing in wealth and developing along large lines. This territory is an empire which will in ten or twenty years more support millions where it now supports thousands."[29]

Opposite page top: The Thomson Dam on the St. Louis River was completed by the Great Northern Power Company in 1907. At the time one of the largest hydroelectric facilities in the world, the Thomson Dam provided electric power to Duluth and nearby areas of St. Louis County. NEMHC S2386 B4f21

The Marshall Wells home office on Duluth's Lake Avenue early in the twentieth century. Marshall Wells was one of a number of wholesale hardware and grocery distribution firms that received goods from the Lower Lakes by boat and delivered them to hardware and grocery stores in Minnesota, Iowa, the Dakotas, and Montana. NEMHC S2386 B11f9

MARSHALL-WE
HEAV
WELLS-STONE MERCANTILE COMPANY
WHOLESALE GROC

A New Century

chapter four

Fire was an ever-present threat in St. Louis County during the years bracketing the turn of the twentieth century. Many of the residences and commercial buildings were constructed of wood, and the use of fireplaces and wood stoves for heat—and candles and gaslights for illumination—almost dictated that the county would be visited by fire each year.

The town of Virginia reported three major fires in the 1890s and early 1900s. Few of the county's urban areas were spared at least one major conflagration each year. When Range communities were chartered in the 1890s, among the first and most important municipal functions attended to were the organization and staffing of a volunteer fire department.

Another threat lurked just outside the city limits. St. Louis County in the years between 1890 and 1920 was covered by a vast, arboreal forest that stretched from the forested hills lining the St. Louis River Valley west and south of Duluth to the Canadian border north of Ely. Loggers had been cutting the white pine of St. Louis County since the 1870s, and the cut accelerated dramatically during the 1890s and early 1900s. Unlike today, when the loggers left a site, there was little cleanup or attention to potential fire danger. Huge piles of slash, bark, and limbs that had been stripped from the marketable logs, were left at the landing to dry out and decompose.

Fire was an ever-present threat in St. Louis County during the years bracketing…

The result was a potential tragedy waiting to happen. The Lake states had been victimized by fire since time immemorial. Fire was nature's way of removing old growth and preparing the way for renewal of the forest.

One of the worst regional fires took place in October 1871, but was all but unknown because it happened at the same time as the Chicago Fire. The Peshtigo Fire burned

The November blow of 1905 wrecked dozens of ships on western Lake Superior, including the freighter Mataafa, *which broke apart off the outer piers of the Duluth Ship Canal. Half the ship's crew, marooned in the stern section, froze to death within sight of downtown Duluth.* NEMHC S2386 B46f11

much of northeastern Wisconsin. An estimated fifteen hundred people died in the fire, which was centered around the logging community of Peshtigo, Wisconsin. In little more than an hour, Peshtigo, which reported a population of two thousand in the 1870 census, was reduced to ash and cinders.[1]

Just over two decades later, Minnesota's turn came. As strikers from Eugene V. Debs' American Railway Union battled with rail police and National Guard troops protecting Pullman cars in the summer of 1894, much of Minnesota baked during one of the hottest and driest summers in years. On September 1, the bone-dry conditions caught fire in at least four locations surrounding Hinckley, halfway between the Twin Cities and Duluth. Before the holocaust had run its course, winds in excess of eighty miles per hour had devastated a three-hundred-square-mile area of Pine County. An estimated four hundred settlers died in the blaze, and the town of Hinckley was destroyed.[2]

St. Louis County was initiated into the roster of communities shattered by forest fires in September 1908 when a fast-moving conflagration engulfed the town of Chisholm. Sitting along the shores of Longyear Lake and described as one of the prettier communities on the Mesabi Range, Chisholm boasted an electric light plant; two newspapers; *The Herald* and *The Tribune;* and two banks. In less than two hours on September 5, 1908, the town of Chisholm was all but gutted by a forest fire that started just northwest of town.

One eleven-year-old girl recalled that "the smoke was so dense that one could hardly see, and the wind nearly took us off

Gilbert sat astride the Duluth, Missabe and Northern tracks halfway between McKinley and Eveleth. Like other mine locations in the area, much of the housing in the community was erected by the mining companies and rented to the miners. Construction of churches and schools quickly followed. NEMHC S2386 B17f1

Onlookers gawk at a 1902 bank explosion in Eveleth. At the time, as is visible from the rubble, 90 percent of the commercial buildings in the Range towns were erected of wood, which made fire an ever-present danger. NEMHC S2386 B16f3

our feet. One lady, running down the street, carried a bird cage. The bottom had fallen out, and the bird had escaped, but she clung to the cage."[3]

When the smoke cleared, the village of six thousand people was in ruins. The business district was gone, and only a handful of houses on the outskirts of the community were still standing. Unlike earlier conflagrations in the Lake states, the 1908 blaze was accompanied by enough advance warning and was confined to a limited enough area that casualties were kept to a surprising minimum.

But when residents returned to what had been the site of their homes, they were faced with the daunting task of rebuilding their community and their lives. Relief trains arrived almost immediately from Duluth and other Range communities, and a canvas city soon surrounded the Chisholm school, the only brick building to survive the fire.[4] Local militia from the nearby villages and mine locations provided security and policing during September and October, and a relief committee began soliciting funds and in-kind contributions for the reconstruction of the town.

Like the mythical Phoenix, Chisholm rose from the ashes. By July 1909, the town had been entirely rebuilt, with many of the buildings in the commercial district encased in fireproof brick. Sidewalks flanked the graded main street, and a new municipal water and sewer system served the town. Best of all, the presence of abundant employment in the region's iron mining industry attracted new residents to the rebuilt community. When the census takers visited Chisholm in 1910, they found eight thousand people living there, 25 percent more than had been in residence the day of the fire.[5]

Forest fires continued to plague Minnesota in the early decades of the twentieth century, with 1910 a particularly troublesome year. In October of that year, the twin villages of Baudette and Spooner on the Rainy River some seventy-five miles north and west of St. Louis County's northwest corner were destroyed in a mammoth blaze that raged along both sides of the United States–Canadian border. When the fire had burned itself out on October 7, 1910, rescuers faced the horrific task of identifying and burying the forty-two known victims of the fire.[6] Others were undoubtedly dead in the trackless bogs south of the Rainy River, but their bodies were never found.

The White Pine Frontier

Residents of the county, however, were more than willing to take the chance of dealing with a forest fire, given the economic boost that logging and forestry gave to the area. In 1900, the State of Minnesota published statistics that showed timber companies operating in St. Louis, Itasca, and Beltrami Counties—the three major counties containing significant stands of white pine—employed almost sixteen thousand workmen in more than three hundred camps.[7]

The camps across St. Louis County were located in the deep woods and linked to the mills in Duluth, Virginia, and International Falls by a spiderweb network of single-track logging railroads. The logging camp concept created many of

The immigrant groups who made up the vast majority of the population of the Range towns in the early 1900s took every chance to celebrate their Americanism. The Fourth of July was a favorite holiday. The annual parade—as shown here in Eveleth early in the century—was invariably followed by a picnic featuring pails of Fitger's Beer, Duluth's favorite beverage. NEMHC S2386 B16f3

By the late 1910s, the city of Hibbing perched on the edge of the Hull Rust Mahoning pit. The existence of high-grade iron ore beneath the city would necessitate the actual removal of most of Hibbing's commercial buildings to a new site in the 1920s. NEMHC S2386 B17f7

Aurora, on the eastern end of the Mesabi Iron Range, was platted in 1903. Within walking distance were a number of mines, including the Mohawk, Miller, Meadow, Fowler, Donora, Perkins, St. James, Weed, and Stephens Mines. It wasn't until 1910, however, that the community lost the look of a frontier mining camp. NEMHC S2386 B13f17

the small communities in northern St. Louis County. Cusson was a good example. Founded in 1913 as a field, or railroad, camp, Cusson was located just north of Pelican Lake along the right-of-way of the Duluth, Winnipeg and Pacific Railway.

The Virginia and Rainy Lake Lumber Company was the Oliver Iron Mining Company of the county's early twentieth century forest resources industry. Established in 1909 as a merger of lumber companies in Virginia, International Falls and Duluth,

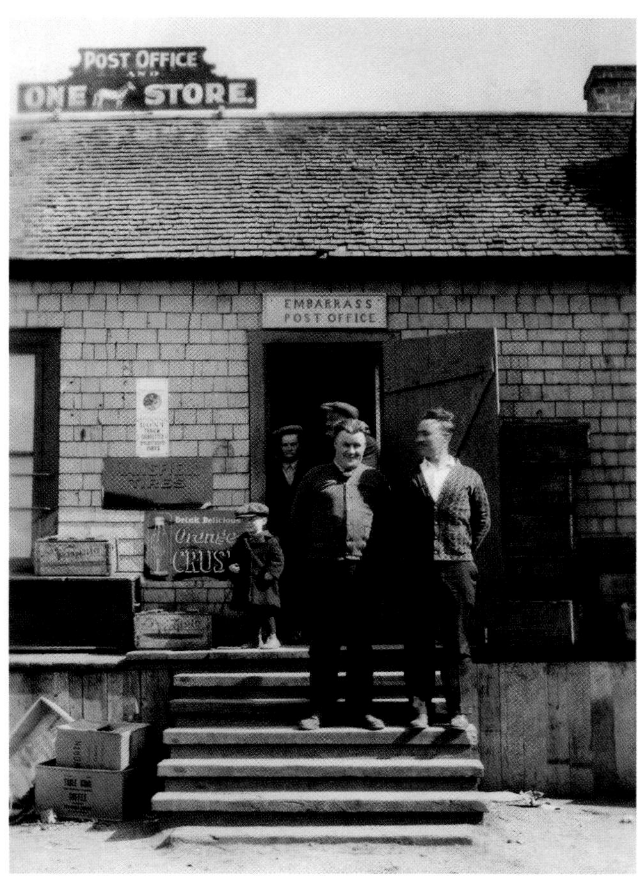

Embarrass was one of the fast-growing agricultural centers in St. Louis County. Founded in 1905 the community and surrounding township of the same name had a population of 49 people at the turn of the twentieth century. Ten years later, the population had increased to nearly 650 people, nearly all engaged in farming. The Embarrass Post Office, seen here in the late 1910s, served rural patrons in the Embarrass River Valley. NEMHC S2386 B16f2

the Virginia and Rainy Lake operated the largest pine sawmill in the world at Virginia on the Mesabi Range and built several of the lumber camps that later became forest villages. In 1910, at the peak of the county's early twentieth century logging boom, the Virginia and Rainy Lake was handling an astounding 300 million board feet of white pine sawtimber a year.[8]

The logging camp at Cusson cost the Virginia and Rainy Lake $83,000 to build. The camp boasted a headquarters building, boiler house, machine and railroad shops, sheds, a pump house, warehouses, an icehouse, and a coal dock. The camp at Cusson also included living quarters for families and single loggers, which also necessitated construction of a schoolhouse, movie theater, doctor's office, general store, and recreation building.[9]

In the 1910s, the Virginia and Rainy Lake operated twelve to fifteen lumber camps scattered across northern St. Louis and adjacent counties. In a typical year at the time, the firm employed nearly 3,000 workers in its camps, as well as several thousand additional employees at mills in Virginia and Duluth. Each of the Virginia and Rainy Lake camps employed 150 to 200 men, including a foreman, clerk, checker, scaler, cook crew, filer, several blacksmiths, watchmen, and horse attendants. The woods crew typically consisted of 125 to 150 individuals.[10]

In the woods, the lumberjacks earned between $2 and $2.50 a day. They worked ten to twelve hours a day, felling white pines with six-foot, two-men saws and limbing the logs with axes. The logs were dropped at the landing where they were stacked and hauled to rail sidings by teams of horses or oxen.

The timber cut in St. Louis County disappeared almost as quickly as it had grown. In 1900, nearly 1 million tons of timber crossed the docks at Duluth-Superior. A decade later, the peak of the cut in the northern third of the county, the volume of lumber shipped through the Twin Ports had slipped to 428,000 tons.[11] More than 1.6 million tons went down the Lake in 1918 to fuel the industrial push of America's participation in World War I. But the next year, 1919, the lumber float from the Head of the Lakes plunged to 94,000 tons.[12] In the early 1920s, the mills that once dominated the Duluth-Superior waterfront fell silent.

Opening the County to Agriculture

The decline of the logging industry in the 1910s created a volume of cleared land suitable enough to attract those interested in agricultural pursuits. Since much of the county was either heavily forested or what would today be classified as wetlands, agriculture came relatively late to the county's industrial mix.

Farming dates back to the county's earliest days, primarily because the horses and mules used in the timber and iron ore industries consumed prodigious amounts of hay and silage. As early as the 1890s, farmers in the southern reaches of the county and on the immediate fringe of the Mesabi Range communities were growing the native meadow grasses—primarily blue joint or redtop—for feeding the region's livestock. But farming in the early days of St. Louis County was a laborious, backbreaking task that involved clearing and harvesting timber, grubbing out stumps, removing rocks, and breaking sod. Even though hundreds of thousands of acres were logged off in the county between 1890 and 1920, a forty-acre farm was considered a fairly large undertaking.

Ironically, the forest fires that swept across St. Louis County in the first two decades of the twentieth century created an ash enriched soil that produced excellent crops, at least during the first four or five years after the fire. Ditch grading, the process of dredging the bogs in the northern half of the county and piling the spoils up in the middle of parallel ditches for road construction, helped drain lands for farming.

Biwabik, five miles west of Aurora, was nineteen years old when this photo was taken in 1911. The community was in the process of building a new water and sewage system that year, and a new two-story hospital replaced the clinic that had burned down in 1905. The city's First National Bank had been founded in 1907. NEMHC S2386 B13f20

In 1890, while the Merritts were still exploring what would become the Mesabi Range, there were just over three hundred farms in the county, most in the southern third of the county. By 1920, there were forty-two hundred farms in St. Louis County, a fourteen-fold increase in thirty years.[13] Similarly, acreage in farms in the county increased six-fold between 1900 and 1920, from sixty thousand acres to more than four hundred thousand acres.[14] But at less than one hundred acres, the average farm was small.

By the time World War I broke out, farmers in St. Louis County were growing timothy and clover, often mixed. Alfalfa was another popular silage crop. The hay crops typically were harvested in late July or August. In the early days, hay was harvested with a scythe, dropped in a swathe and then hand-forked into shocks, which were usually later combined into stacks for fall curing. Later, mechanization—first with horse-drawn equipment and later with tractors—allowed the county's farmers to bring in more hay with fewer hands.

The weather limited the hay crop. Farmers in northern Iowa and southern Minnesota frequently took two hay crops off

Top: Many of the farming communities of northern St. Louis County were populated by Finnish immigrants. The Finnish community in early twentieth century Minnesota was often fractured along political and social lines. "Good Finns" attended the Suomi Synod or Finnish Evangelical Lutheran Church, belonged to the kalevala *temperance society and voted Republican. "Bad Finns" slept in on Sunday, read* Tyomies*—The Workman—and belonged to the IWW.* NEMHC S2386 B16f2

Buyck in far northern St. Louis County was named for settler Charles Buyck, who petitioned the County Board of Commissioners to organize a township in 1906. At the time there were fifteen people living in the proposed township, which one reporter later called "practically wild land." NEMHC S2386 B13f23

their fields in an average year. But the frost-free season in northern St. Louis County was in the neighborhood of one hundred days. And during the wet years of the 1910s, it was frequently touch-and-go whether area farmers would be able to get one hay crop off their fields before winter set in or the shoots rotted. Still, the value of hay and forage planted in the county increased more than 400 percent just between 1910 and 1920, from $300,000 to $1.375 million.[15]

Much of the silage was consumed locally by the county's livestock. Most farmers kept at least several head of cattle, both for beef and dairy products. In the northern reaches of the county, a number of farmers herded sheep, which were remarkably hardy and able to forage from spring until the first snowfall. Sheep, however, often fell prey to carnivores, primarily timber wolves and coyotes, called brush wolves by the local residents. For most of the twentieth century, St. Louis County paid residents a bounty for each wolf they killed until aircraft hunting of wolves in the 1940s and 1950s landed the county's wolves on the endangered species list.

Hogs were less popular than sheep, although prized for the hams and bacon they provided. Nearly every farm in the county counted at least a small flock of chickens for eggs and meat, and a number of farms prided themselves on the fat geese they raised.

As the county opened up in the 1910s, the acreage planted to cereal crops increased dramatically, although from an essentially nonexistent base prior to 1900. Because of their hardiness, oats and rye were favorites among St. Louis County farmers. Since little of the

Iron Range workers socialized with their fellow immigrants at mining company picnics during the summer. Here, children enjoy getting together at an Oliver Iron Mining picnic in Virginia just after World War I. NEMHC S2386 B29f6

soil in the county was sandy, potatoes did not do as well as they did further west in Koochiching, Lake of the Woods and Beltrami Counties. St. Louis County's Finnish farmers, however, could always be counted upon to find a spot in their gardens for turnips and rutabagas. And even urban country residents often planted a vegetable garden behind the house and canned the county's abundant wild berry crop—usually blueberries, wild strawberries, raspberries, and thimbleberries—in the late summer and early fall.

The 1910s were relatively dry, warm years, and commodity prices were the highest they had ever been during World War I. As a result, the total value of all crops in St. Louis County hit nearly $4 million in 1920, four times the value of crops in 1910 and twice the value of crops in the county at any time during the next twenty years.[16]

"Arise, Ye Prisoners of Starvation . . ."

By the time, the Mesabi Range celebrated its twenty-fifth anniversary in 1917, the United States was at war, and the flood of ore from the Minnesota ranges had reached truly gargantuan proportions. In 1916, 1917, and 1918, as America prepared to get involved in the war that had embroiled Europe since 1914, production from the Mesabi Range was double what it had been the year that Archduke Franz Ferdinand was assassinated at Sarajevo. In those three years ending in 1918—the year the Armistice was signed in the Compeigne Forest—Duluth-Superior handled an average of 38 million tons of Mesabi Range ore each year.

The war had special meaning for many of the immigrant workers digging ore for the Oliver Iron Mining Company and other Range mining firms. A not insignificant number of South Slav miners returned to their Balkan homeland in the 1900s and 1910s to fight the Ottoman Turk—or their Serb, Croat, Slovene, or Bosnian neighbors—in the seemingly never ending series of wars that wracked the peninsula between 1900 and 1914.

For the miners of the Mesabi Range, there also was evidence that Oliver Iron Mining Company brooked no opposition in its quest to wrest the red ore from the earth of St. Louis County. When an estimated ten to twenty thousand workers,

Workers at the North Star Mine in 1913 prepare for a shift. Miners going to work changed clothes in the "dry," and typically walked to the work site in the pit. In 1916, miners striking against Oliver Iron Mining Company demanded to be paid "portal to portal," which meant that they wanted to be paid from the time they arrived at the dry rather than at the pit. NEMHC S2386 B29f5A

the vast majority of them immigrants, walked off their jobs at Mesabi Range iron mines in June 1916, the company quickly imported armed guards from the "detective agencies" that supplied strikebreakers to companies facing strikes by workers.[17]

What the miners were already calling "the steel trust" refused to meet with its disaffected workers. Oliver cited the radical nature of Industrial Workers of the World (IWW) organizers who were leading the strike and quickly asked Governor J. A. A. Burnquist to nationalize the Minnesota National Guard for strike duty on the Mesabi Range. Ironically, much of the state's National Guard force, including units from Chisholm, were already on duty, assisting General John "Blackjack" Pershing in tracking down Pancho Villa's rebels along the Mexican border.

The IWW had run well-publicized strikes of millworkers in Lawrence, Massachusetts, in 1912 and silkworkers in Paterson, New Jersey, in 1913. In the days before federal and state labor-management negotiation laws, there were few rules. Strikes in mining districts were often violent, because miners knew and understood how to handle dynamite, residents of isolated mining districts grew up using guns for hunting, and mining companies weren't hesitant to hire armies of "detectives," or gunmen, to enforce labor peace. There was violence during the IWW strike on the Mesabi. At least one miner, one county deputy sheriff, and an innocent bystander were shot and killed during strike-based confrontations; federal and state labor investigators later blamed most of the violence on mine guards and police.[18]

Three years before, a Western Federation of Miners (WFM) strike against the Calumet and Hecla Copper Company in Michigan's nearby Keweenaw Peninsula claimed the lives of a half-dozen miners, strikebreakers, and armed company guards. The strike ended tragically in December when a WFM Christmas Party at Calumet's Italian Hall was brought to an abrupt close by a shout of "fire." Before order could be restored, nearly seventy women and children piled up dead on the staircase from the second floor to the street.[19] Folk artist Woody Guthrie wrote one of his most haunting songs about the incident; descendants still claim that company armed guards raised the fatal call of fire.

For the IWW, the strike was a necessary battle in the class war against capitalism. Big Bill Haywood, the general secretary of the IWW, was familiar with the Range from the short-lived and unsuccessful 1907 Federation strike against Oliver. For the 1916 strike, Haywood sent in his best organizers, including Carlo Tresca, an Italian anarchist who would be murdered by Mafia hitmen in New York in 1942 for his criticism of Italian dictator Benito Mussolini, and Elizabeth

The economic boom in timber and mining during the early part of the twentieth century created a home-grown group of millionaires in Duluth and on the Mesabi Iron Range. Marshall H. Alworth first came to Duluth from New York in 1873. For the remainder of the 1800s, Alworth invested in timberlands in St. Louis County. When the Merritts brought in the Mesabi Range in 1892, Alworth was already heavily invested in mining lands. With associates W. H. Cole, W. C. Agnew, and Chester Congdon, Alworth made another fortune in the iron mining industry in the years prior to World War I. In 1909, he built the Alworth Building in downtown Duluth, the first skyscraper north of the Twin Cities. NEMHC Portrait File

Gurley Flynn, the movement's twenty-five-year-old "rebel girl," who would later assist Range native Gus Hall in running the Communist Party USA. Frank Little, the tough IWW organizer who would be lynched by mine guards in Montana the next year, spent much of the summer of 1916 on the Iron Range.[20]

Although the strike petered out in the late summer of 1916, and Oliver and the other iron mining companies would not recognize their workers' right to organize for another generation, the miners got most of what they had originally gone on strike for. By the end of the year, wages had risen nearly 20 percent and hours were reduced from a twelve-hour-day to an eight-hour-day.[21]

By 1915, many of the mines on the Mesabi were already several hundred feet deep. Completion of a high-voltage electric transmission line from the Thomson Dam on the St. Louis River to the Mesabi Range in 1913 allowed for the electrification of a number of mines in the region in the years before World War I. One of the first tasks that engineers harnessed electric power for in the open pit mines of the Mesabi was pumping excess water out of the bottom of the pits. Roleff, NEMHC S2386 B29f5

Oliver and its parent, U.S. Steel Corporation, had been exposed to adverse publicity because of the strike, and the companies needed labor peace to cope with a surging demand for steel to fuel America's war efforts. Coupled with a labor shortage on the Range, the rising demand for steel ushered in an era of comparative labor-management peace.

The Fire Next Time

The labor shortage was equally severe in Duluth during World War I. The Twin Ports was one of the nation's fastest-growing communities in the late 1910s. U.S. Steel's Minnesota Steel Company had cast its first ingot in late 1915 at a state-of-the-art mill announced in 1907. U.S. Steel had invested more than $20 million to build the mill with six open-hearth furnaces, byproduct coke ovens, and adjacent Atlas Portland Cement mill.[22] To attract workers to the site located on the far west end of the Duluth metropolitan area, U.S. Steel built Morgan Park, a planned community of homes and retail businesses within walking distance of the new mill.[23]

Open pit mining in the first two decades of the twentieth century was done by steam shovels, like this Marion at a mine north of Eveleth about 1910. NEMHC S2386 B28f7

In 1917 and 1918, even U.S. Steel had difficulties attracting workers to the relatively high-paying jobs at the steel mill and cement plant. The American entry into World War I had been a shot in the arm for the Twin Ports' moribund shipbuilding industry. The McDougall-Duluth shipyard at Riverside and the Globe and Superior Shipbuilding Company yards on the Wisconsin side of the harbor employed more than ten thousand people in 1917 and 1918 building merchant ships for America's allies.[24] Before shipbuilding wound down in 1920, the three yards at the Head of the Lakes would build seventy-five vessels.[25]

Less than a month before World War I ended in November 1918, Duluth, southern St. Louis County, and adjacent areas to the south had shared the fate of Peshtigo in 1871, Hinckley in 1894, and Chisholm in 1908. A mammoth forest fire ravaged the area between Moose Lake and Duluth on October 12, 1918. The daylong fire followed a hot, dry summer and was fanned by winds in excess of thirty miles per hour. Fires south of Moose Lake and west of Brookston coalesced in mid-afternoon, wiped out Cloquet, and finally burned out that night in the townships surrounding Duluth.

The fire curled around the city and actually caused more damage on the city's east end than in neighborhoods west of downtown. The blaze damaged hundreds of houses in Lakeside, Woodland, Lester Park, and on Rice Lake Road. One of the casualties was the city's Northland Country Club, which a reporter described as "a charred heap of ruins."[26]

The final death toll for what became known as the Cloquet Fire or the Moose Lake Fire was more than five hundred people.[27] The St. Louis River communities of Floodwood and Brookston in the far southern end of the county were all but destroyed, and refugees from as far away as Moose Lake streamed into Duluth that awful Saturday night.

St. Luke's and St. Mary's—the city's two trauma care hospitals—were already at capacity treating victims of the rapidly spreading Spanish Influenza pandemic. The day before the fire, Duluth Mayor Clarence R. Magney declared a state of emergency because of the outbreak, which had stricken eighty people in the city in little more than ten days.[28]

Nevertheless, the two hospitals made room for the injured of the fire. The crowding at the hospitals was alleviated on October 16 when the city's hospitals and the Duluth Public Health Department opened an emergency influenza hospital downtown. The next day, St. Mary's opened an emergency ward at the hospital for influenza victims.[29]

The 1908 and 1918 fires aside, the 1900s and 1910s were golden years for St. Louis County. The steady growth of employment in the county's iron ore, timber, and maritime commerce industries meant that there were jobs for everyone who wanted them. County population had doubled from 1900 to 1910 and grew by an additional 25 percent from 1910 to just over 206,000 people in 1920.[30] In the decade ahead, the county and its people would grow the infrastructure needed to take advantage of advances in transportation and communications.

A steam shovel at the Mahoning Mine near Hibbing about 1895 strips overburden and loads dirt and ore into railroad cars. Coal for the operation of the steam shovels was hauled from the docks at Duluth-Superior in otherwise empty ore cars. NEMHC S2386 B28f4

Bicycles were all the rage in the 1890s, and tinkerers affixed small, primitive gasoline piston engines to create the ancestor of today's motorcycle. In 1909, the first race held at the Indianapolis Motor Speedway was a motorized bicycle race. By 1920, motorized bicycles frequently were raced at mining company picnics on the Iron Range. NEMHC S2386 B29f6

Augustus B. Wolvin was one of the most influential people in the nation's iron ore industry in the first quarter of the twentieth century. Named the first manager of U.S. Steel's Pittsburgh Steamship Company subsidiary in 1901, Wolvin left the steel trust three years later to design five-hundred-foot vessels for his own Acme Steamship Company. Wolvin's vessels set the standard for the Great Lakes ore trade for the next thirty-five years. NEMHC Portrait File

Mines in the early twentieth century were extremely labor-intensive and employed hundreds of workers to handle tasks that machinery would perform in another generation. Here, an oiler at the Leonidas Mine quenches his thirst at a water fountain on the Mine's 340-foot level. NEMHC S2386 B28f1

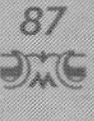

By 1910, all of the iron ore dug from the Mesabi Range came from open pits. The only deep-shaft mining in St. Louis County was at mines such as the Soudan, Chandler, and Pioneer on the Vermilion Range. Here, miners prepare to ride the man car to a stope deep within the mine. Note the skip in front of the man car, which brought ore to the surface and served as a counterweight to the man car.
NEMHC S2386 B28f8

An aerial view of Morgan Park and the Minnesota Steel Company mill and cement plant about 1916. U.S. Steel used the Duluth plant to supply barbed wire to the open range livestock industry of the Dakotas, Montana, and Wyoming. The Atlas Portland Cement mill adjacent to the steel mill produced high quality cement, which found a ready market all over the Upper Midwest. NEMHC S2386 B14f21

Work at deep-shaft mines differed radically from ripping ore out of an open-pit mine. Miners used compressed air drills to bore holes into the rock face holding the ore body. They then tamped the drill holes with explosives and fired the charge. Muckers followed, loading the rock and ore into cars for the trip to the surface. Until about 1910, crews in most deep-shaft mines used two-man drills. Ingersoll-Rand then introduced the one-man drill, which most miners hated because it meant working in solitude instead of with a partner. The miners called the drill "the widow-maker" and feared that if anything happened, they wouldn't be found until the end of their shift. NEMHC S2386 B28f9

When an estimated twenty thousand miners walked off the job in June 1916, Victor L. Power was the popular mayor of Hibbing. Although Power initially sympathized with the mining companies, he came to believe that the striking miners had legitimate grievances and that the companies' intransigence was prolonging the strike. Van Brunt, v. 3, p. 970

The Duluth Aerial Bridge has been the signature structure for the Twin Ports since it was first erected in 1905. During the ten-month navigation season, the bridge operated twenty-four hours a day. The man car ferried passengers back and forth across the Duluth Ship Canal and even had space for a streetcar and a fire truck. NEMHC S2386 B1f8

In the early days of the twentieth century, there was a clear class division in the employment ranks of Mesabi Range mining companies. Engineers, superintendents and white-collar employees typically were Anglo-Saxons or North Europeans, descendants of immigrants who had arrived in America in the mid-nineteenth century. The miners and laborers were Finn, Slav, and South European immigrants who had arrived in Minnesota only recently. NEMHC S2386 B28f7

Baseball was a passion that united all Iron Rangers, whether native born or "just off the boat" immigrants. Here, a Hibbing National Guard unit's baseball team relaxes between innings at a game played at Camp Cody, New Mexico, probably during the 1916 campaign against Mexican rebel Pancho Villa. NEMHC S2386 B34f18

The office staff at the Oliver Iron Mining Company gathered for a group photograph in March 1906. NEMHC S2386 B29f6

Mining location residents took pride in the appearance of their homes and gardens. Here, children of residents of mining company housing in Hibbing show off their families' vegetable gardens about 1915. NEMHC S2386 B17f13

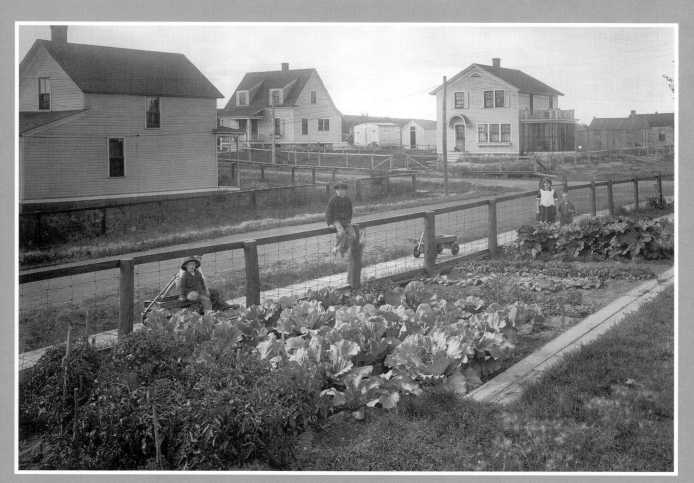

The mining company paternalism that some miners found objectionable was more often than not eagerly embraced by other employees. Oliver Iron Mining Company operated recreational facilities for its workers, including this electrically lighted billiards room in Hibbing in 1913. NEMHC S2386 B17f11

Living in St. Louis County provided residents bountiful outdoor recreational opportunities. Here, a resident traverses Trout Lake near Ely on a summer morning in 1910. NEMHC S2386 B16f1

Top: As early as the turn of the twentieth century, residents of the Vermilion Range were discovering that Burntside Lake near Ely was a wonderful place to picnic, fish, and pick wild blueberries and raspberries. NEMHC S2386 B16f1

World War I gave Mesabi Range residents the chance to prove their allegiance to their adopted country. This infantry company from Eveleth included soldiers of Finnish, Italian, Serb, Croat, and Bulgarian extraction. Ironically, more than a few of the members of the unit had relatives who were fighting for the Austro-Hungarian Empire, Imperial Germany's ally in the war. NEMHC S2386 B16f3

Clarence R. Magney was mayor of Duluth in the fall of 1918 when the Head of the Lakes was hit with the double whammy of forest fire and Spanish influenza. Magney went on to a distinguished career as a St. Louis County Court judge during the 1920s and 1930s. NEMHC Portrait File

RIVERSIDE REVIEW

MAY, 1918

COPYRIGHT 1918 BY

M^c^DOUGALL-DULUTH CO.
DULUTH, MINNESOTA

The Riverside Review *was a monthly employee communication distributed during World War I to the several thousand shipyard employees of McDougall-Duluth Company in Riverside.* NEMHC Serials

Iron Range worker housing was leased to employees, most of whom were first-generation immigrants who took inordinate pride in their new homes. NEMHC S2386 B17f3

*The Duluth shipping firm's formal name was the Minnesota Atlantic Transit Company, but everybody on the Great Lakes knew the company's vessels—*Ace, King, Queen, Jack, *and* Ten*—as the Poker Fleet.* McKenzie, NEMHC S2422 no. 15388

The Oliver Club of U.S. Steel Corporation's Oliver Iron Mining Company subsidiary included a library, meeting rooms, exercise facilities, and game rooms for Iron Range employees of the firm. NEMHC S2386 B29f6

Depression and War

chapter five

In the 25 years following the 1918 armistice ending World War I, the world indelibly changed. The armistice at Compiegne ended what historians call "the long nineteenth century," the 129-year period between the onset of the French Revolution and the end of the "war to end all wars." The armistice ushered in what other historians have called "a return to normalcy," an era that lasted for all but the final two months of the 1920s.

For America—and St. Louis County—the 1920s truly were an uninterrupted period of normalcy. The manufacturing initiative that had made the world safe for democracy during World War I had required millions of tons of steel. Much of the manufacturing capacity was seamlessly transferred to consumer durables in the two years following the war. That translated into continued strong business for the iron mines of the Mesabi Range during the 1920s.

In Detroit, Henry Ford's Rouge Steel was turning out thousands of tons of steel sheet stamped into body parts for the ubiquitous Model T Ford put America on wheels with the Model T, and the automobile entrepreneur boasted that the customer could have the car in any color he wanted, so long as it was black.

In the 25 years following the 1918 armistice ending World War I, the world indelibly…

The explosion of automobiles on the roads meant that dirt and gravel highways were no longer acceptable. St. Louis County was an early pioneer in creating and maintaining all-weather roads. By the late 1920s, U.S. Highway 53 from Duluth north to Virginia and the Iron Range was paved with crushed stone and concrete, as was U.S. Highway 61 from south of Duluth to Two Harbors. U.S. Highway 169 also was paved from Virginia down the Iron Range to Grand Rapids, and U.S. Highway 2 from Grand Rapids to Proctor was paved with crushed stone and a bituminous coating.[1]

The huge Peavey terminal elevator on the Duluth waterfront helped make the Zenith City one of the world's busiest grain ports during the 1920s. NEMHC S3766 B2f13

The Mesabi Iron Range spawned one of the nation's iconic transportation ventures. Andrew "Bus Andy" Anderson was selling cars in Hibbing in the 1910s when he began offering residents scheduled rides in a Hupmobile. In 1915, Anderson partnered with Carl Wickman and several others to form the Mesaba Transportation Company, which provided bus service to the Range communities. By 1919, the company was manufacturing its own buses in Hibbing as Mesaba Motor Company. Wickman moved the company to Duluth in 1922, and two years later, merged with several other companies in Minnesota and Wisconsin to form Northland Transportation Company. By the end of the decade, Northland merged with other bus companies to form the nucleus of Greyhound Corporation.[2]

Some more daring county residents also were taking to the air in the 1920s. On a cold day in late February 1919, pilot Walter Bullock and observer Major William Garis put their biplane down on the ice of Superior Bay. After refueling and repairs at Duluth's Oatka Boat Club, the two pilots took off for the return trip to the Twin Cities. They landed at Pine City and White Bear before touching down on the ice of Lake Calhoun just before 4 p.m. The round-trip flight covered 320 miles and constituted the first of what would be many flights between Duluth, the Iron Range, and Minneapolis.[3] Ten years later, retired mathematics teacher Anne C. Filiatrault became the first woman to solo as a pilot in Duluth.

Cultural, Recreational, and Educational Attainment

The 1920s were a decade of full employment and the creation for the first time of a leisure economy. Many of the county's cultural institutions date from the decade of flappers and bathtub gin, primarily because residents had time to enjoy pursuits outside the workplace.

Most rural people in St. Louis County used the abundant wood resources of the area to heat their homes. A jokester has scrawled "farmers' winter crop" across this photo of a woodpile near Meadowlands in the winter of 1922–1923. NEMHC S2386 B18f1

Winter holds St. Louis County in its grip from November until April. For Duluthians, the first lake vessel through the Ship Canal is as sure a sign of spring as is the first robin farther south. But the first laker into the harbor often has to fight its way through pack ice. NEMHC S3766 B2f5

William E. Culkin, who worked closely with Walter Van Brunt and the Chicago staff of the American Historical Society in the preparation of the magisterial 1921 three-volume history of the county, helped found the St. Louis County Historical Society in 1922 and served as the society's first president. Two years later, in 1924, the Minnesota Arrowhead Association was formed to help promote tourism in St. Louis County and along the North Shore of Lake Superior.

County residents enjoyed unparalleled access to the great outdoors, including boating and fishing in Lake Superior and the many lakes of the wilderness area north of the Iron Ranges. Skiing was increasingly popular during the winter months in the 1920s, especially on the hills surrounding Duluth. The area boasted curling rinks in all of the major Range towns, and Duluthians enjoyed ice boating on Superior and St. Louis Bays. Ice hockey more often than not was played on outdoor rinks.

Baseball was a cultural glue that bound together the families of the immigrant groups who had populated much of the county. When Babe Ruth, the star of the New York Yankees' 1920s dynasty, appeared at Duluth's Lyric Theater in the fall of 1926 as part of a twelve-week vaudeville tour, the King of Swat drew more than ten thousand people in Saturday and Sunday performances. Ruth appeared with Detroit Tiger players Harry Heilman and Heinie Manush. While in the Twin Ports, he had breakfast with Superior resident Dave Bancroft, then the manager of the Boston Braves of the National League.[4]

As the county developed a middle class, its young people could afford to stay in school and participate in organized sporting activities. Because of the climate, St. Louis County developed budding young stars in a number of sports, including skaters Lois Dworshak and Anne Haroldson, and later, Bobby and Ruby Maxon, all of Duluth. Anne Govednik, who would go on to be an Olympic swimming star, received her early training in her hometown of Chisholm.[5] That Iron Range communities such as Chisholm enjoyed such amenities as swimming pools was a function of the prosperity that permeated the Mesabi Range during the 1920s.

Hibbing had physically been moved in 1919 when the Hull-Rust-Mahoning pit encroached on the city's downtown area. Movers spent much of the summer and fall of 1918 and 1919 jacking up buildings and moving them via steam log haulers and tracked vehicles to the Central Addition, about one mile away.[6] The move was paid for by the Oliver Iron Mining Company, which retained the surface mining rights to what had once been the downtown of Old Hibbing. Oliver reportedly paid between $17 million and $20 million to move old buildings to and erect new buildings in Central Addition, called New Hibbing.[7]

Because of the school and property taxes paid by the mining companies in the 1920s, the Iron Range schools were the envy of the entire Midwest. With an assessed valuation of $135 million in 1920, the Hibbing School District—technically Independent School District No. 27 of St. Louis County—benefited from an annual school tax of more than $1.4 million. The school district, which encompassed an enormous area of six entire townships and sections of eight other townships, operated seventeen school buildings in the early 1920s.

The Hibbing Technical and Vocational High School—the district's crown jewel—was completed at a cost of $3.8 million in 1923. The brick and stone school included a

The area around Meadowlands and Toivola boasted sandier soil than most of the rest of the county, allowing region's primarily Finnish farmers to develop a reputation for raising some of the better seed potatoes in the 1920s. NEMHC S2386 b18f1

Because of the immense physical size of St. Louis County, commissioners were forced early on to locate county services in communities other than the county seat of Duluth. In 1910, the county built a $275,000 courthouse in Virginia to serve residents of the northern half of the county. The Virginia courthouse was more than doubled in size during the early 1920s. NEMHC S2386 B20f7

St. Louis County established a reputation in the twentieth century of building some of the most modern and advanced schools in the country. Funded by mining company taxes, the Hibbing School District spent more than $3 million in the early 1920s to build a new high school and junior college. At the time the new high school was completed in 1923, a majority of schoolchildren in Minnesota were still attending one-room schools. NEMHC S2386 B17f7

swimming pool, gymnasiums, shops, meeting rooms, and auditoriums. Residents boasted that it was the finest such facility in the Midwest, if not the nation. When President Calvin Coolidge, who maintained a summer home on the Brule River south of Superior, toured the Iron Range in the summer of 1928, he specifically asked to be conducted through the Hibbing High School.[8]

Not to be outdone by their neighbors in Hibbing, the Virginia School District completed the magnificent new Virginia Technical High School in 1925. NEMHC S2386 B20f7

The Bottom Falls Out

For most of America, one of the watershed dates of the era was October 29, 1929. Pundits called it "Black Tuesday," but in reality it was the beginning of a slow motion stock market crash that carried the nation into the worst economic downturn since the Panic of 1893. The upheaval on Wall Street would plunge America into a depression that wouldn't lift for more than a decade. It would devastate the nation's steel industry, and with it, most of St. Louis County.

The Iron Range economy was seemingly healthy in 1929, except for the final closing of the Virginia and Rainy Lake Lumber Company mill in Virginia, once the world's largest and busiest sawtimber mill.[9] Nineteen twenty-nine, in fact, had been a boom year on the Vermilion and Mesabi Ranges. Nearly 50 million tons of iron ore had gone down to the Lake Superior iron ports aboard the ore cars of the D&IR and the Duluth, Missabe and Northern, the biggest single year for ore shipments out of Duluth since World War I.[10] Employment in the mines had declined, primarily because Oliver Iron Mining Company had invested heavily in electrification during the 1920s.[11] But the U.S. steel industry had broken records in 1929 with production of more than 63 million tons of finished steel.[12] There was little to suggest that the industry's record production would not continue into the future.

The grand opening of the Duluth Aerial Lift Bridge in the spring of 1930 signified the prosperity that had characterized Duluth and St. Louis County ever since the first shipment of Mesabi Range ore had come down the hill nearly four decades before.[13]

When the fall came, it was all the more brutal for its swiftness. The stock market opened lower on Monday morning, October 28, and never looked back. Most of the orders pouring across the exchange floor were sell orders, with blue chip stocks in the majority.[14]

A big part of the problem was psychology. The market's irrational exuberance of the late 1920s had been fueled by brokers and individual investors buying on margin, essentially putting up a small percentage of a price per share and then booking profits when the stock went higher. By Monday, banks all over America were making margin calls, demanding full payment for stock purchases they had loaned money on.[15] Before the day was over, the Dow Jones Industrial Average had dropped more than 38 points on volume of more than 9 million shares.

When the stock markets opened on Bloody Tuesday, the carnage continued. "On that day, shortly after the New York

One of "Bus Andy" Anderson's buses made a scheduled stop in downtown Hibbing in the early 1920s. The thriving metropolis was home to a bustling retail shopping district that served customers from all across the Iron Range. NEMHC S2386 B17f8

Stock Exchange opened," historians Charles A. and Mary R. Beard wrote, "the bottom almost fell out of the buyers' market. Only sellers appeared in force, and they dumped hundreds and thousands of shares into this well-nigh bottomless pit at any prices that were offered. Down, down, down dropped prices as the throng of brokers milled and shouted around the posts of the specialists who were supposed to have buyers for all comers. Out in the streets early editions of newspapers announced the calamity."[16]

Before the day was over, the Dow had dropped nearly another 30 points as 16.4 million shares changed hands—often at whatever price the market would bear.[17] The economy continued to worsen into the new year.

At U.S. Steel, the largest steel company in the country and the parent of Oliver Iron Mining Company, 1930 production was off by a quarter from 1929, and profits were cut in half. Shipments from the Mesabi Range also were down a quarter, and in the spring of 1931, a number of mines elected not to open for the season. Pickands-Mather's big Mahoning pit near Hibbing was idled for the first time in thirty years.[18]

Business executives tried to make the best of the situation. Minnesota Power and Light Company, a part of the giant

Electric Bond and Share Company, heeded the call of President Herbert Hoover for construction projects to put Americans back to work. MP&L built a $3.5 million steam electric station in West Duluth—later named for company president M. L. Hibbard—on the bayfront in 1930 and 1931.[19]

Nothing anyone did or tried, however, masked the fact that the region's economy was in freefall. The Great Lakes shipping season wrapped up in November 1931. When the spring of 1932 came, fewer than half of the Great Lakes bulk fleet of four hundred vessels fitted out for the season, and nearly half of those vessels laid up early.[20]

In 1932, the Mesabi Range was desolate. Almost none of the mines were open, and those that were operated with a skeleton staff. Two of every three workers on the Mesabi Range were laid off for extended periods from 1932 through the middle of the decade. Banks failed. Those residents of the county who made a living from farming watched as commodity prices spiraled downward. By the fall of 1932, wheat prices in Minnesota were thirty-six cents a bushel, a quarter of what they had been seven years earlier.[21]

The reason for the devastation on the northern St. Louis County Iron Ranges was simple. From Pennsylvania's Monongahela Valley to South Chicago, and from Birmingham, Alabama, to Ohio's Mahoning Valley, the U.S. steel industry was dead and dying that Depression winter. Mill capacity skidded to 20 percent. Most industry observers were quietly predicting that the steel companies would hunker down and pretty much write off 1932. In the end, the steel companies chalked up 1933, 1934, and 1935 as lost years, too.

People on the Iron Range and in Duluth made do, and not a few county residents gardened, fished, hunted, heated their homes with wood, and raised a few dairy and beef cattle, hogs, and poultry. Because the vast majority of the homes on the Range were owner-occupied, the homelessness that afflicted the big cities wasn't particularly evident in Hibbing, Virginia, Mountain Iron, and other Range mining locations. Unlike in cities like Chicago and New York, most people in St. Louis County kept their homes. A sizable number of Iron Rangers leased homes from the mining companies, particularly in the more isolated locations. Rather than let the homes sit vacant,

Unlike communities in other parts of Minnesota, Hibbing's location was somewhat of a movable feast. During the 1910s and 1920s, parts of Hibbing were physically moved from one part of town to allow the mining companies to unearth the rich ore deposits beneath the community. Here a tracked vehicle moves a two-story frame house from the city's Kerr Location on November 3, 1924. NEMHC S2386 B17 f8a

most of the mining companies established a moratorium on evictions during the early 1930s.

The election of Franklin Delano Roosevelt as president and Floyd B. Olson as governor in 1932 alleviated some of the worst excesses of the Depression. To halt the erosion of confidence in the monetary system, FDR instituted a bank holiday in March 1933. His New Deal administration quickly began putting people back to work on federal public works programs like the Works Progress Administration (WPA) and the Civilian Conservation Corps (CCC). Thousands of young men from St. Louis and surrounding counties worked at CCC camps in places like Brimson and Bemidji.

One of the parks that Samuel F. Snively oversaw development of was Duluth's Enger Park. The juxtaposition of the automobile culture and the leisure class is amply illustrated by this 1927 shot of the parking lot at the Enger Park Golf Course. NEMHC S2386 B18f11

The highlight of the summer for many Rangers was the company picnic. Here, the women test their endurance in a game of tug-of-war at the Oliver Iron Mining Company Picnic in the summer of 1925. NEMHC S2386 B29f6

But even the weather seemed to conspire against northern Minnesota during the Depression. 1936 was a year of extremes. Temperatures plunged to the -40s and even -50s for weeks at a time across the northern half of the county during the brutally cold winter. Even Duluth, normally warmer during the winter months because of its proximity to Lake Superior, experienced three weeks of sub-zero temperatures in January and February.[22]

By summer, the parched Northland was broiling in the fiercest heat wave in recorded history. On July 14, Duluth and the Mesabi Iron Range reported temperatures above 105 degrees, an all-time record.[23] The tenth day of a two-week heat wave covered most of the Midwest like a wool blanket.[24] Residents of the northeastern part of the county anxiously scanned the skies to the east to ascertain if a big forest fire blazing out of control near Tofte in Cook County would burn west into St. Louis County. Some six hundred CCC enrollees were enlisted to fight the worsening fire.[25]

"A New Day Is At Birth"

The massive loss of jobs triggered labor militancy in the region unprecedented in the county's history. Many of the area's immigrant workers had fled oppression abroad, and they found solidarity in unions and radical political movements in their new country. When the region's economy hit bottom during the Depression, workers were quick to hit the streets.

The first to demonstrate their unhappiness with prevailing conditions were the longshoremen of the Twin Ports in 1931. With thousands of residents unemployed, the packet freight companies that hired the bulk of the longshoremen could reduce wages with impunity. In the spring of 1931, the International Longshoremen's Association (ILA), which had represented as many as three thousand workers in the Twin Ports during World War I, began a series of work stoppages on the waterfront.

County residents loved baseball during the middle years of the twentieth century. Here the municipal baseball stadium in Chisholm sits empty after a Saturday afternoon game in the 1920s. Chisholm was the home of Dr. Archibald "Moonlight" Graham, who played one game in the Major Leagues. Graham later became famous to a generation of Americans when writer Joseph Kinsella used Graham's character in the novel Shoeless Joe. *Actor Burt Lancaster played Graham in* Field of Dreams, *the movie made from Kinsella's novel.* NEMHC S2386 B14f8

Employers responded by hiring strikebreakers, and flying squads of longshoremen, led by Duluth organizer E. L. "Buster" Slaughter, fell upon the strikebreakers with baseball bats and bricks. Duluth police joined the melee on the side of the companies, and the battles continued for much of the spring.[26]

Teamsters in Duluth and on the Range organized during 1933. While most of the activity was peaceful in the Northland, Teamster organizing efforts erupted in Minneapolis in the spring of 1934. Before the labor strife was over, four Teamsters were shot dead in the streets of Minneapolis by police and armed company guards. One Duluth Teamster remembered riding shotgun down Hennepin Avenue to protect his fellow Northland Teamsters from violence.[27]

On the Mesabi Range, the Farmer-Labor Party was in its ascendancy during the 1930s. The Minnesota Federation of Labor had run candidates on the Farmer-Labor Party label as early as 1918, and William Leighton Carss was elected to the U.S. Congress from Duluth and St. Louis County in 1918 on an independent labor ticket. The Farmer-Labor movement formed on the Range in 1923 to oppose Victor Power for another term as mayor of Hibbing, and it helped elect Floyd B. Olson as governor in the 1930s. When the popular Olson died in office, his successor, Elmer Benson, announced his bid for re-election in 1936. Benson was swept into office on the strength of a 70 percent vote majority in St. Louis County outside of Duluth. The Farmer-Labor landslide also elected John T. Bernard to the Eighth Congressional District seat.[28]

Known as the "singing fireman," Bernard was an Eveleth-based organizer for John L. Lewis' Congress of Industrial Organizations (CIO) in the mid-1930s. Ironically, the CIO and the Steelworkers Organizing Committee (SWOC), the predecessor of the United Steelworkers of America (USWA), made little headway with Range iron workers during the 1930s. Many of the unemployed had left the Range in the early 1930s, and by 1935, production had resumed at a number of the area's mines. It would not be until the World War II years that the steelworkers were organized at Mesabi and Vermilion Range mines.[29]

The passenger vessel North American *departs Duluth beneath the aerial bridge sometime in the early 1920s. The* North American *and her near identical twin sister, the* South American, *were two of dozens of elegant cruise vessels that operated scheduled service across the Great Lakes in the late nineteenth and early twentieth centuries.* NEMHC S2386 B43f9

Bernard broke with the New Deal administration over the arms embargo of the Spanish Republic during the civil war in Spain. Meanwhile, the Farmer-Labor Party and the CIO split with the American Federation of Labor (AFL). The AFL endorsed the Republican candidate in the 1938 congressional election, and Bernard was defeated after one term in Congress, although the Minnesota Federation of Labor did endorse Benson for another term.[30]

Perhaps the most militant group of workers in St. Louis County in the 1930s was the lumberjacks. The "jacks" cut pulpwood for the newsprint mills that had replaced the white pine sawtimber industry in the 1920s. Most of the lumberjacks worked in camps, cutting pulp all winter and paying the timber agents room and board. Gus Hall, who worked the camps in the 1920s before going to Moscow, later recalled that conditions in the camps were difficult. The food, Hall said, was inedible, and the beds were full of bugs.[31] Life in the camps was nothing like the romantic picture often painted of the lumberjack era.

Many of the loggers were Finns, and a substantial number, like Hall, were Communists. Trouble flared up in the winter of 1937, and it appeared the camps might go on strike. But the AFL intervened and sent organizers from the Carpenters and Joiners Union to negotiate a contract with the timber companies.

Most of the timber operators were essentially small businesses who were not well organized themselves. They simply ran a camp or two and sold the pulp to the newsprint mills.[32] The need to organize did force the timber operators into forming the Minnesota Timber Producers Association (MTPA), an organization that exists nearly seventy years later.

Led by Ilmari Koivunen, a pulp cutter from north of Mountain Iron, more than two thousand lumberjacks unleashed a series of wildcat strikes across most of northeastern Minnesota, northwestern Wisconsin, and into the Upper Peninsula of Michigan.[33] There was scattered violence across Michigan's U.P., but for the most part, the strikers were peaceful in Minnesota. They did dominate the headlines in November 1937, however, when they blocked highways leading to the Northland on the first weekend of deer season. Minnesota Governor Elmer Benson had been working behind the scenes to get the two parties together. When he couldn't get to his cabin on the North Shore because of the roadblocks, Benson took back roads to Duluth, met with the strikers and the timber companies and demanded that the two sides settle the strike that weekend.[34]

Labor problems and radical politics were quickly set aside on Sunday, December 7, 1941, when St. Louis County and the rest of America reacted with shock and horror to the Japanese sneak attack on the U.S. Navy base at Pearl Harbor, Hawaii. The United States was at war, and the Iron Ranges and Duluth collectively rolled up their sleeves to win the battle for production.

Mobilizing for War

Martin Kuusisto was typical of the county residents who helped America win World War II. The son of Finnish immigrants, he was born in Duluth in 1910 and served in the Minnesota National Guard as a teen-ager during the late 1920s. Like many Finns, Kuusisto's politics were decidedly left of center, and he was one of the organizers for the timber workers' strike of 1937. He left before the strike ended to join the International Brigades fighting for the Republic in Spain. Kuusisto survived a torpedo attack off Barcelona, and was assigned to a French artillery brigade. He participated in several campaigns before returning to the United States in early 1939.[35]

Kuusisto was drafted for active duty in the U.S. Army in 1941 and was sent to Officer Candidate School because of his combat artillery experience in Spain. On June 9, 1944, he

Mountain Iron in 1930 was one of the Mesabi Range communities that had sprung up in the short span of thirty-eight years since the Merritts started shipping ore from the Missabe Mountain Mine in 1892. NEMHC S2386 B18f5

In 1930, there were nearly five thousand farms in St. Louis County. Nearly 80 percent of the county's crops consisted of hay and forage, which was fed as silage to dairy cattle. The cooperative movement was popular among the county's rural residents. This farmer-owned creamery marketed milk and butter on the Mesabi Iron Range and in Duluth during the 1930s. NEMHC S2386 B20f8

landed at Utah Beach on Normandy, on D+3. During the next ten months, he participated in some of the fiercest fighting on the Western Front, at the Falaise Gap, the Battle for Paris, the Huertgen Forest, and the Battle of the Bulge.[36]

Thousands of county residents, no matter their nationality, politics, religion, or station in life, served their country during World War II. Some, like Duluth's Michael Colallilo, were decorated with the nation's highest military honors for bravery in combat. Many others spent the war in stateside posts. Too many never came home, dying instead in places that most Americans had never heard of before 1941: Midway Island, Bataan, the Kasserine Pass, Schweinfurt, Tarawa, Anzio, Bastogne, Luzon, and Okinawa.

Those on the home front contributed mightily to Allied victory during the war. The Mesabi and Vermilion Iron Ranges and the Lake Superior iron ports were where the arsenal of democracy truly began. Oliver Iron Mining Company, Pickands-Mather, M. A. Hanna, and the other iron mining companies on the Vermilion and Mesabi Ranges began digging ore as soon as Congress answered President Roosevelt's call for war on December 8, 1941. Before the 1942 season closed almost exactly a year later, 92 million tons of iron ore had gone downbound from Lake Superior to the steel mills lining the Lower Lakes. More than two-thirds of that production came from the Mesabi and Vermilion Ranges.[37]

The flood of Minnesota iron ore charged the blast furnaces that provided hot metal for the finishing mills. Minnesota ore produced the raw steel that made sheet and plate and coil that went into Sherman tanks, battleships, Dodge trucks, Browning machine gun barrels, and all of the implements of war that turned the American military into the best-equipped battle force in the world. The Iron Ranges shipped more than 50 million tons in 1943, 1944, and 1945. All told, the Mesabi and Vermilion Ranges shipped more than 200 million tons of ore during World War II, two-thirds of the iron ore consumed by the American steel industry from 1942 to 1945.[38]

St. Louis County contributed to the war effort in another concrete way. As in World War I, the Twin Ports of Duluth and Superior were a beehive of activity. Zenith Dredge, Marine Iron and Shipbuilding, Walter Butler Shipbuilders, Barnes Duluth Shipbuilding, Globe Shipbuilding, and Lake Superior Shipbuilding built more than two hundred vessels between 1941 and 1945. At the crest of shipbuilding in 1944, more than fourteen thousand people were employed in the bustling shipyards on both sides of the harbor.[39]

Among the vessels coming down the ways, or launch ramps, in the Twin Ports during the war were thirty-eight "180s," multipurpose buoy tenders that the U.S. Coast Guard operated all over the world for the next sixty years.[40] The Duluth shipyards also built dozens of cargo vessels for the U.S. merchant marine, as well as for America's British allies. The shipyards launched seagoing tugs, submarine tenders, Navy frigates, and coastal freighters.

Those left behind on St. Louis County's home front did their part to win the war. Civilians conserved gasoline, sugar, coffee, and dozens of other commodities. They rolled bandages for the Red Cross and scoured backyards and alleys from Ely to Floodwood searching for ferrous and nonferrous scrap that could be melted down for the war effort. They purchased millions of dollars worth of war bonds at work and tended Victory gardens behind neatly kept houses in Biwabik and Buhl and Lester Park.

Perhaps the biggest change wrought by World War II in St. Louis County—and the nation as well—was society's realization that women were as well suited to handle most jobs as their male counterparts. Hundreds of county women saw service in the armed forces, including the WACs, WAVEs, SPARs, and WASPs. Women from every community in St. Louis County answered the call to work when labor shortages in the mines and shipyards became critical after 1942; in the Duluth and Superior shipyards, women were reputed to be better welders than men. More thousands of young women from the Northland moved to the Twin Cities and worked in defense industries, such as the huge ammunition depot at New Brighton.

The atomic attacks on Hiroshima and Nagasaki in early August 1945 brought an end to the war and ushered in a dangerous new world in which St. Louis County would continue to play an important role. But there were concerns on the Mesabi Iron Range the fall the war ended that the inexhaustible supply of iron ore was finite after all.

Most communities in St. Louis County that boasted a sizable Finnish population sported a Kalevala Hall. The Kalevala Hall was part social club and part temperance society and was an integral part of community life for the region's first- and second-generation Finns in the 1920s and 1930s. NEMHC S2386 B20f7

Unemployed workers lined up on Michigan Street in Duluth on a winter morning in the early 1930s. After 1932, both the state and the federal government offered public works programs to put county residents back to work. The New Deal Works Progress Administration spent nearly $10 million on payroll in Duluth from 1933 to 1939 and was responsible for building everything from an expansion of the community's airport to a new sewer system downtown. NEMHC S2386 B21f6a

A prize guernsey herd near Floodwood, about 1930. Notice the relatively new barn and the attached silo for storing hay and feed for the herd. St. Louis County dairy farmers benefited from federal dairy supports that established Eau Claire, Wisconsin, as the base point in the United States for shipping milk products. NEMHC S2386 B31Af11

By the end of the 1920s, St. Louis County farmers were increasingly using mechanized equipment to plant and harvest crops. This crew on the W. F. Haenke farm near Gilbert in 1928 was pulling a horse-drawn combine to harvest potatoes. Because of the small size of area farms, relatively few county farmers at the time could justify the cost of a John Deere or Fordson tractor. NEMHC S2386 B31Af11

COMPANY 719, C. C. C.

Brimson, Minn.

CAMP CHARLES

'By the Waters of Lake Sullivan'

March 10, 1934

Company 719, C.C.C. was organized on May 26, 1933, at Fort Snelling, Minnesota. Grovener C. Charles, 1st Lieut., 3rd Inf. was the first commanding officer, assisted by L. W. Kehe, 2nd Lieut. Eng-Res. On May 26th, 16 enrollees joined the organization and on May 27th, 178 more enrollees joined, bringing the total strength to 194 men. Our Commander, Lieut. Charles, was replaced by A. M. Shearer, Capt. S. C. on June 3rd. On the evening of June 10th the Company proceeded from Fort Snelling to their permanent station, State Camp S-51, located at Brimson, Minnesota. The camp site itself is 11 miles west of Brimson on the shores of Lake Sullivan. We immediately christened it Camp Charles in honor of our first Company Commander. Long hours of hard work, consisting of stump-removing, leveling and general cleaning up, transformed a rather rough and widerness-like site into a very orderly and presentable Camp. On June 24th, 20 Local Experienced men from St. Louis County joined the Organization. Louis M. Rosenbladt, 1st Lieut. Med-Res. joined us on June 29th. At this time our Forestry Supervisors consisted of Mr. Donnelly, Camp Superintendent, and J. C. Ryan, local Forest Patrolman. Olin G. Reiniger, Capt. Ing.-Res. joined the Company on July 31st but stayed with us only a short time, being relieved August 22nd. John H. Hettinger, Capt. 13th Cavalry, took over the Command from Capt. Shearer on August 12th. October 1st began a new enlistment period—about 130 of the men re-enlisting. Also on that date, Fred Gardner, Jr., 1st Lieut. FA-Res. replaced Capt Hettinger as Commander. It was about this time that the Company moved into their permanent barracks. The buildings are of palisade construction. The Camp has an appearance of a quadrangle, six barracks on each side, each of them 20 ft. by 40 ft., 18 men to the barracks, with the exception of one end barrack which is larger and consists of the Orderly Room in the front part and the Supply Room in the back with a small room between which serves as quarters for the Headquarters men. On one end of the quadrangle is the Recreation Hall, a building 20 ft. by 80 ft, consisting of a barber shop, canteen, pool table, ping-pong table, newspaper and magazine rack, together with a number of small tables and benches where the men can sit down and write, read, etc. At the other end of the quadrangle we have our Mess Hall, a building 36 ft. by 130 ft. with the Officer's Mess attached. On one side of the inclosed area we have the hospital and on the other, the Officer's quarters. The Forestry Service office and garages are off to one side about 200 yards away from the Camp proper. Mr. A. W. Nelson succeeded Mr. Donnelly as Camp Superintendent on October 23, 1933.

On November 3rd, 63 recruits from Kansas joined the Organization bringing our number up to the authorized strength of 212. On November 5th, Otto C. Person, Capt. CA-Res. took over the Command from Lieut. Gardner, although the latter remained with the Company until December 11th, being replaced at that time by Roland Brooks, 1st Lieut. FA-Res. On November 23rd, Lieut. Kehe was relieved. Our Camp Surgeon, Lieut. Rosenbladt was relieved from duty on December 21st and was replaced on January 5th by Philip E. Gordon, M. D., Contact Surgeon. On January 10, 1934, 11 Local Experienced Men from St. Louis County joined the Organization bringing the strength up to 207 men. On January 31st, Sgt. William H. Perry was relieved from further duty with the Company at his own request. Sgt. John E. Brink, the only remaining Regular Army man, was relieved on February 14, 1934. Our Camp Educational Adviser, John Westerlund, joined the Organization on March 7th.

Camp Charles was located on Lake Sullivan, eleven miles west of Brimson. The camp's official name was Company 719, CCC, and it was formed in the spring of 1933 at Fort Snelling. Most of the original complement of recruits were from the local area, although sixty-three recruits arrived from Kansas in November 1933 to bring the strength of the company above two hundred men. NEMHC S4568

Five CCC recruits at Camp Charles near Brimson inspect a grouse chick one of the men has discovered. St. Louis County young men from every walk of life got their first chance at employment in the CCC during the Depression. The exposure to a spartan, communal lifestyle in the CCC made the transition to military life during World War II a relatively simple matter for most veterans of the camps. NEMHC S4568

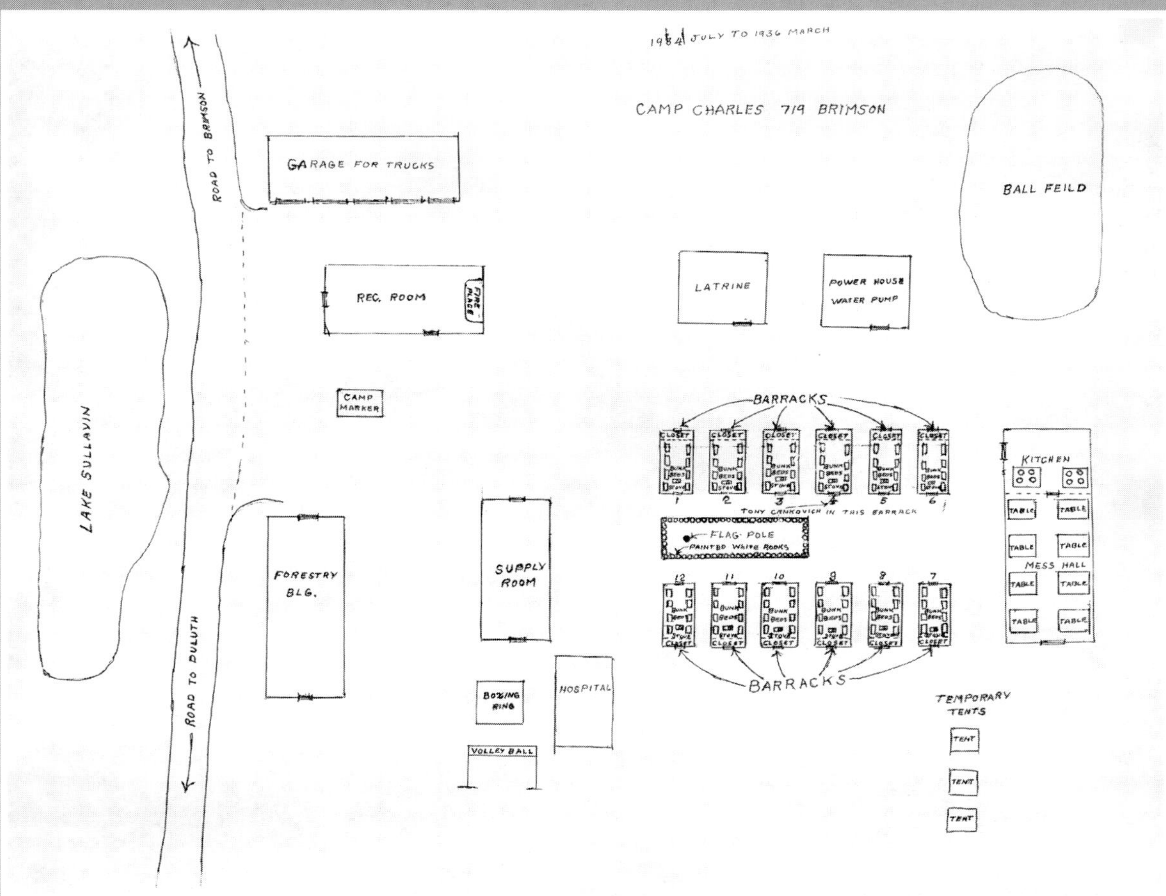

A sketch map of Camp Charles in 1934 revealed a sizable facility with barracks for nearly two hundred men, a kitchen, recreational facilities, a medical clinic, a garage and forestry center. The entire facility was illuminated with electric lights, supplied by a generator in the powerhouse. NEMHC S4568

Another federal program that had a large footprint in St. Louis County was the Works Progress Administration, and its sister program, the Public Works Administration. WPA and PWA financed several thousand projects in the county and put thousands of county men to work at a time when the mines, docks, and factories were either shut down or on reduced hours. One example of the WPA impact on St. Louis County is the Hibbing Memorial Building, erected in 1935. NEMHC S2386 B17f7

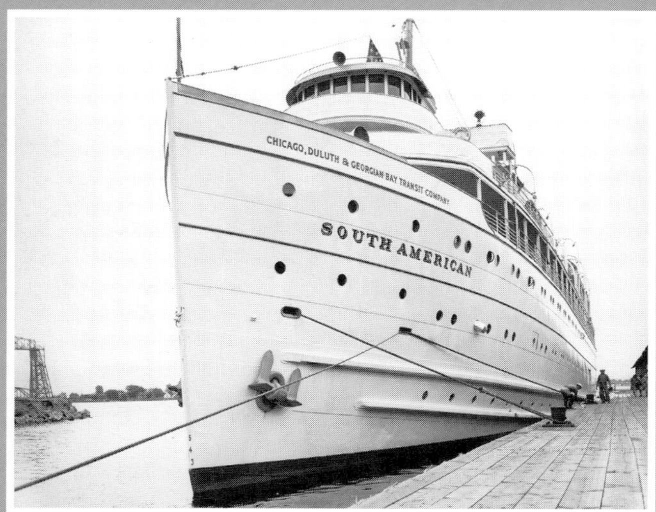

Duluth in the 1930s was the Upper Lakes terminus for a Great Lakes cruise industry that carried passengers in comfort and style to Chicago, Detroit, Cleveland, Buffalo, and dozens of other Great Lakes communities. The South American *was one of the flagship vessels of the Chicago, Duluth and Georgian Bay Transit Company and was a regular caller on the Twin Ports. She sailed on a weekly basis between Duluth and Buffalo and had room for more than five hundred passengers.* NEMHC S2386 B43f13

By the mid-1930s, the small bus companies that "Bus Andy" Anderson and Carl Wickman founded on the Mesabi Range in the years just before the outbreak of World War I had matured into the Greyhound Corporation. During the Depression, the nationwide system of inter-city buses was increasingly proving to be a tough competitor to the passenger railroads. NEMHC S4443f1

Top: One of the ironies of the Great Depression was that the 75 percent to 80 percent of people who had jobs were relatively well off. Prices actually declined for much of the 1930s, which increased the purchasing power of consumers. Shops like these at the busy corner of Lake Avenue and Superior Street in Duluth were often busy with shoppers even during the darkest days of 1932 and 1933. NEMHC S3766 B2f6

Leisure activities became more common as the economy moved to a forty-hour workweek under labor legislation championed by the Roosevelt administration. Here, Duluthians play tennis at Chester Park. NEMHC S2386 B18f8

The Noronic, *seen here departing the Duluth Ship Canal on an August evening in the mid-1930s, was the queen of the Canadian Lakes. Built at Port Arthur in 1913, the* Noronic *was a frequent visitor to Duluth for the next third-of-a-century. In 1949, her wooden superstructure caught fire in a Toronto berth; 119 passengers died in the conflagration, which caused U.S. and Canadian maritime regulators to place tough new restrictions on Great Lakes cruise vessels.* NEMHC S2386 B43f9

Radio was another development of the 1920s and 1930s that shrunk the global community. WEBC Radio was fifteen years old in 1939 when comedian Bob Hope and Dorothy Lamour appeared in the station's NBC studio to promote their latest On the Road *movie.* NEMHC S2386 B329f10

The iron mining plants of the 1930s were sophisticated operations that included rock crushers, screening systems, and washing plants. The washing plants were necessitated by the depletion of the highest-grade iron deposits during the 1910s and 1920s. By the time this photo was taken in Virginia in 1939, the industry was well on the way to recovery from the ravages of the Depression. NEMHC S2386 B28f9

The mechanization of open-pit iron mining during the 1930s and 1940s essentially meant that many of the jobs lost in the first wave of layoffs after the onset of the Depression in 1932 and 1933 were not replaced when the mines resumed production in the late 1930s. Here, an electric shovel loads a haul truck at a mine near Virginia in 1939. NEMHC S2386 B28f9

The size of the pits on the Mesabi Range was almost beyond description. One writer once called them Minnesota's version of the Grand Canyon, and work crews were kept busy year-round laying railroad track and haul roads to expedite the flow of ore from mining sites within the pits. NEMHC *S2386 B28f9*

Left: In 1939, Crown Prince Olav of Norway and his consort, Princess Martha, visited Duluth and dedicated the Enger Memorial Tower overlooking Enger Park. The Prince paid tribute to Bert J. Enger, the Norwegian immigrant and Duluth millionaire whose $50,000 gift to his adopted city made possible Duluth's purchase of the land that became Enger Park. Crown Prince Olav told more than five thousand spectators that he was honored to take part in the ceremony dedicating the tower named for a fellow countryman. NEMHC S3766 B1f29

Above: Superior native Ernie Nevers thrilled the crowds during the 1920s when he played with the Duluth Eskimos, a charter member of the National Football League. The Eskimos left Duluth in 1929 for New Jersey and eventually became the Washington Redskins franchise. Nevers went on to coach the NFL's Chicago Cardinals from 1929 to 1931 and again during 1939. NEMHC S2386 B35 f7

Top: Less than two months after the Crown Prince of Norway dedicated Enger Tower, Duluth officials began tearing down another Twin Ports landmark. The Duluth Incline Railway, which had climbed the hillside at the foot of Seventh Avenue West every day since 1891, was quietly closed in the late summer of 1939 following a $20,000 loss in its last year of operation. The Incline was scrapped that fall, and much of the scrap steel found its way to blast furnaces for the nation's war effort less than two years later. NEMHC S3766 B2f9

The United States' entry into global war in December 1941 changed the American economy almost overnight. St. Louis County's children did their part for the war effort by participating in scrap drives. This group of Morgan Park schoolchildren spent afternoons and weekends in the fall of 1942 scouring their neighborhoods for scrap metal, aluminum, copper, and paper that could be recycled for ammunition and equipment needed by American troops on the far-flung battle fronts of the world. NEMHC S3007 B2f56

Every institution in St. Louis County pitched in to help win World War II. St. Mary's Hospital provided a site for the storage of three hundred units of frozen blood plasma, an advancement from the early 1940s that allowed hospitals and blood banks to store blood for years at a time. Gallagher, NEMHC S2386 B7f20

Saint Mary's Hospital

SISTERS OF SAINT BENEDICT

OFFICE OF THE
HOSPITAL ADMINISTRATOR

DULUTH, MINNESOTA
January 12, 1942

entire report notice!

Mrs. D. W. Wheeler, Chairman
Women's Volunteer Service
Duluth Office of Civilian Defense
200 Alworth Building
Duluth, Minnesota

Dear Mrs. Wheeler:

After your telephone call, I got busy and investigated the opportunities for the group of women whom you stated want to work from two to five hours a week making surgical sponges. We will be very happy to have these women make up stick and tonsil sponges for the operating room, as well as to wrap Obstetrical pads each of which is wrapped individually in a tissue paper napkin.

As regards the students from the Junior College who wish to work once a week for two hours, we also would have sufficient work for them. There are many swabs to be made and various types of packs and other surgical supplies. Since the hospital uses many newspapers and magazines for wrapping up discarded dressings, etc., also for other purposes, we would appreciate having these workers bring us clean newspapers and assist in cutting them to the various sizes ready for use.

Will you please have whoever is responsible for these groups report to Miss Evelyn Chaillee at our hospital. For the present she is to direct this work. Assuring you of our appreciation, I am

Sincerely,

Sister Patricia, OSB

Sister M. Patricia, O.S.B.
Administrator

SP;ck

Duncan Town sorority from Jr. College assigned to above

OPA FORM R-1705 A
FORM APPROVED BUDGET BUREAU No. 08-R321

UNITED STATES OF AMERICA
OFFICE OF PRICE ADMINISTRATION
SHOE PURCHASE CERTIFICATE

Nº 19242

CHECK ONE [ ] FOR EMPLOYERS, INSTITUTIONS FURNISHING (NOT SELLING) SHOES [ ] FOR RESALE BY A MEMBER OF THE TRADE DATE

10—33001-1 ☆

NAME

AMOUNT IN FIGURES

ADDRESS

IS HEREBY AUTHORIZED TO BUY

AMOUNT IN WORDS

PAIRS OF SHOES

TYPE OF SHOES AUTHORIZED OR WORDS "UNRESTRICTED TYPE SELECTION"

ISSUING OFFICE (CHECK ONE) [ ] DISTRICT [ ] STATE | CITY AND STATE | SIGNATURE OF ISSUING OFFICER | TITLE

THIS CERTIFICATE is not valid unless signed by an issuing officer authorized by the Office of Price Administration. It may be used to purchase shoes no later than 30 days after the Date of Issue and is not valid for deposit at a bank after 60 days from the Date of Issue.

JUL 3 1947

ONE UNIT | ONE UNIT | ONE UNIT | ONE UNIT | ONE UNIT | ONE UNIT | ONE UNIT | ONE UNIT | ONE UNIT

Form OPA R-504

BUY UNITED STATES WAR BONDS AND STAMPS

UNITED STATES OF AMERICA
OFFICE OF PRICE ADMINISTRATION
GASOLINE RATION CARD
Nº 3825773

B-3

THE ACCEPTANCE AND USE OF THIS CARD CONSTITUTE AN AGREEMENT THAT THE HOLDER WILL OBSERVE THE RULES AND REGULATIONS GOVERNING GASOLINE RATIONING AS ISSUED BY THE OFFICE OF PRICE ADMINISTRATION

OWNER'S NAME

STREET ADDRESS

CITY OR POST OFFICE STATE

MAKE BODY STYLE

VEHICLE REGISTRATION NO. STATE OF REGISTRATION

(SIGNATURE)

READ INSTRUCTIONS ON REVERSE SIDE OF THIS CARD

JUL 3 194

ONE UNIT | ONE UNIT | ONE UNIT | ONE UNIT | ONE UNIT | ONE UNIT | ONE UNIT | ONE UNIT | ONE UNIT | ONE UNIT

St. Mary's Hospital coped with the growing shortage of nurses during the war by enlisting the help of senior Girl Scouts and students from the Duluth Junior College. Many of the young women who volunteered at St. Mary's, St. Luke's, and other county hospitals during the war went on to practice nursing for a career. NEMHC S3007 B2f33

Rationing intruded upon every facet of life during World War II. County residents needed ration books to purchase shoes, meat, tobacco, coffee, sugar, alcohol, and just about every other commodity imaginable. NEMHC S4474 B6f4

Perhaps the biggest sacrifice St. Louis County residents made during World War II was gasoline rationing. A society that had gotten used to driving anywhere in northern Minnesota on all-weather roads, even during the Depression years of the 1930s, was restricted to using several gallons of gasoline a week. Because the Japanese had overrun most of the world's rubber plantations in Southeast Asia, tires also were tightly rationed throughout the war. NEMHC S4474 B6f4

Launchings in Twin Ports shipyards took place as frequently as every other day during the summers of 1942 and 1943. Probably the most famous launching happened in May 1943 when the Dionne Quintuplets of Quebec appeared in Superior to help launch a merchant vessel and to sell war bonds. NEMHC S3766 B2f5

Top Left: The Woodrush *was one of the "180s" that came down the ways at Duluth shipyards during World War II. The "Woody," as she was known to her crews, tended buoys in Lake Superior out of Duluth from 1944 to 1978. She proceeded to the East Coast where she was outfitted for saltwater duty. The* Woodrush *then sailed to Sitka, Alaska, where she was stationed from 1980 to her decommissioning in 2000.* NEMHC S2386 B31f7

The M.V. Titusville *was one of more than two hundred vessels launched by Twin Ports shipyards between 1942 and 1945.* NEMHC S2386 B44f1

The coal docks lining the Duluth-Superior harbor were an integral part of the region's war effort, handling coal shipped up the Great Lakes from Pennsylvania and Ohio and destined for the steel furnaces of the Twin Ports and the Mesabi Range. Ironically, the Middle Eastern oil supplies that were opened up by the war spelled the end for much of the Great Lakes coal trade in the decade following the war. NEMHC S3766 B2f5

Top Left: The "Rosie the Riveter" image of American women during World War II was particularly appropriate in St. Louis County. Because of the county's mining and manufacturing economy, many women found work in traditional male pursuits. Here, a female employee welds wheels at the Duluth, Missabe and Iron Range Railway shops in Proctor during the spring of 1944. Zweifel-Roleff, NEMHC S3744

Samuel F. Snively dominated Duluth politics for most of a generation. Snively was elected mayor in 1921 and was re-elected three times during the next sixteen years. Snively was a tireless promoter of the city's municipal parks and oversaw completion of Skyline Boulevard from Fond du Lac nearly twenty-five miles to the Lester River. After his defeat for a fifth term in 1937 at the age of seventy-seven, Snively went to work for the city as a part-time parks supervisor. He was still working for the city parks when he died in 1952 at the age of ninety-two. NEMHC Portrait File

The fact that some female residents of the county found work on the factory floor during the war did not negate the reality that many area women continued to work in more traditional retail careers. This photo of the staff of Hugo Freimuth's Dress Shop in Duluth in 1945 demonstrates that St. Louis County women of all ages were part of the work force. NEMHC S3766 B2f7

In the 1920s, the Hull Rust Mine near Hibbing was the most productive iron ore pit on earth. It was iron ore from the Hull Rust and dozens of other open-pit mines across the Mesabi Range that allowed the United States to begin its march to world prominence in the Great War. NEMHC S2386 B27f22

Duluthian Julius Barnes carved a name for himself on the national scene during the 1920s. Barnes had helped bankroll Alexander McDougall's shipyard during World War I and then teamed up with Herbert Hoover to help feed Europe in the years after the war. He spent much of the 1930s and 1940s lobbying Congress to join with Canada to build the St. Lawrence Seaway and was the major investor in one of the Twin Ports' largest shipyards during World War II. Barnes died in Duluth in 1959, two and a half weeks before the first ocean-going vessel arrived at the Head of the Lakes via the St. Lawrence Seaway. NEMHC S3025 B12f2

The Clubhouse at Chester Park in the 1930s. The Skyline Parkway above Duluth connected a number of the city's parks during the early years of the twentieth century. NEMHC S2386 B18f8

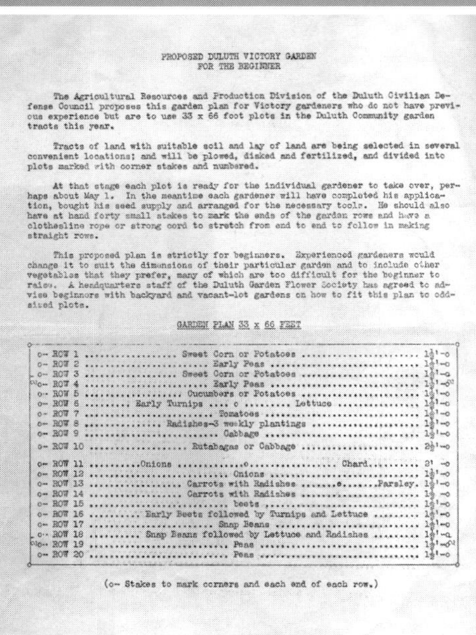

PROPOSED DULUTH VICTORY GARDEN
FOR THE BEGINNER

The Agricultural Resources and Production Division of the Duluth Civilian Defense Council proposes this garden plan for Victory gardeners who do not have previous experience but are to use 33 x 66 foot plots in the Duluth Community garden tracts this year.

Tracts of land with suitable soil and lay of land are being selected in several convenient locations; and will be plowed, disked and fertilized, and divided into plots marked with corner stakes and numbered.

At that stage each plot is ready for the individual gardener to take over, perhaps about May 1. In the meantime each gardener will have completed his application, bought his seed supply and arranged for the necessary tools. He should also have at hand forty small stakes to mark the ends of the garden rows and have a clothesline rope or strong cord to stretch from end to end to follow in making straight rows.

This proposed plan is strictly for beginners. Experienced gardeners would change it to suit the dimensions of their particular garden and to include other vegetables that they prefer, many of which are too difficult for the beginner to raise. A headquarters staff of the Duluth Garden Flower Society has agreed to advise beginners with backyard and vacant-lot gardens on how to fit this plan to odd-sized plots.

GARDEN PLAN 33 x 66 FEET

Row	Planting	Width
o- ROW 1	Sweet Corn or Potatoes	1½' -o
o- ROW 2	Early Peas	1½' -o
o- ROW 3	Sweet Corn or Potatoes	1½' -o
o- ROW 4	Early Peas	1½' -o
o- ROW 5	Cucumbers or Potatoes	1½' -o
o- ROW 6	Early Turnips .... o .... Lettuce	1½' -o
o- ROW 7	Tomatoes	1½' -o
o- ROW 8	Radishes-3 weekly plantings	1½' -o
o- ROW 9	Cabbage	1½' -o
o- ROW 10	Rutabagas or Cabbage	2½' -o
o- ROW 11	Onions ....o.............. Chard	2' -o
o- ROW 12	Onions	1½' -o
o- ROW 13	Carrots with Radishes ....o.......Parsley.	1½' -o
o- ROW 14	Carrots with Radishes	1½ -o
o- ROW 15	beets	1½' -o
o- ROW 16	Early Beets followed by Turnips and Lettuce	1½' -o
o- ROW 17	Snap Beans	1½' -o
o- ROW 18	Snap Beans followed by Lettuce and Radishes	1½' -o
o- ROW 19	Peas	1½' -o
o- ROW 20	Peas	1½' -o

(o- Stakes to mark corners and each end of each row.)

Instructions for planting a Victory garden during World War II. NEMHC S3007 B2

During the first half of the twentieth century, Duluth-Superior was one of the nations major coal ports. Heating coal was shipped up the Great Lakes from Ohio and West Virginia to the Twin Ports and then distributed by rail to coal dealers located as far away as Montana. Gallagher, NEMHC S2386 B28f9

A steam shovel loads ore cars on the Mesabi Range in 1943, part of a deluge of iron ore that helped turn America into the Arsenal of Democracy during World War II. Roleff, NEMHC S3021 B16f50

1956 Interlude

Nobody really understood in 1956 that a big hunk of the future of American popular culture was encapsulated in a skinny high school sophomore in Hibbing on St. Louis County's Mesabi Iron Range.

Dwight D. Eisenhower was elected to his second term as president that year, and Americans worried things were perhaps a little too normal. The book *The Man in the Gray Flannel Suit* accurately described the long-term shift from a blue-collar society to a white collar culture of 9 to 5 workdays, a couple of Big 3 automobiles in each garage, and cookie cutter houses in the suburbs.

That was much more a perception than a reality on the Mesabi Range in 1956. Most working folks still made their blue-collar living hewing iron ore from the ground and sending it down the Lakes to the steel plants in Chicago and Pittsburgh. All those huge Chryslers, Lincoln-Mercurys, and Buicks consumed a ton or more of steel apiece, which translated into good jobs for people in Hibbing, Virginia, Eveleth, Chisholm, and Duluth.

The suburban development that characterized growth in the Twin Cities beginning in the mid-1950s never really took hold in Duluth and the Range towns of St. Louis County. Instead, the 1950s saw the beginning of a decades-long trend when homeowners moved to large acre-plus lots in the rural townships surrounding Virginia, Hibbing, and Duluth.

One jarring deviation from normalcy in 1956 America was the rise of a teen-age music culture fueled by the postwar baby boom, the largest age cohort in American history. The music the baby boomers would listen to during the next twenty years was an amalgam of styles with deep roots in American society. There was rock-a-billy music from the hill country of Texas, mournful blues from the Mississippi Delta, doo-wop from the streets of Philadelphia, and gospel from the African-American communities of Chicago and Detroit.

In 1956, the merger of music styles came together in something called rock and roll. It was personalized by Elvis Presley, a scrawny kid from Tupelo, Mississippi, who knew how to shake his hips and drive teen-age girls wild. Presley paved the way for a generation of rock music stars, none more unique and individualistic than Bobby Zimmerman, the fifteen-year-old sophomore at Hibbing High School. Zimmerman was already into the music of Little Richard and the ascetic style of James Dean, the iconic movie actor killed in an automobile accident the previous year.[1] As Bob Dylan, Hibbing's Bobby Zimmerman would revisit Highway 61 and redefine American music.

The End of An Era

If Bobby Zimmerman represented America's future, another resident of St. Louis County in 1956 represented the nation's past. The death of Albert Woolson of Duluth in the first week of August 1956 marked the end of an era. Woolson, 109, had been a teen-age volunteer in the Union Army in 1864. Nine decades later, he was the last of the more than 2 million men who had donned the blue uniform of the Union during the long and bloody Civil War.

Woolson, who was born in New York state the same day in 1847 as Thomas Alva Edison, also was the last survivor of the Grand Army of the Republic, the veterans' organization that dominated American politics during the late nineteenth and early twentieth centuries.[2] He grew up on a farm in upstate New York and was apprenticed as a carpenter by his father, Willard Woolson. The senior Woolson was one of the seventy-five thousand volunteers called up by President Lincoln in the wake of the Confederate attack on Fort Sumter in the spring of 1861; the next spring, he had been seriously wounded at the pivotal Battle of Shiloh on the Tennessee River.

Woolson's father was sent to Minnesota to recover from his wounds, and young Albert, then fifteen, accompanied his mother to what was then the Minnesota frontier. Shortly after they arrived, Willard Woolson died. Albert and his mother remained in Windom through the Sioux uprising of 1863.

When Colonel William Colville, the hero of the First Minnesota at Gettysburg, came back to the state to raise a heavy artillery regiment in 1864, young Albert Woolson signed up. He left Minnesota a private in Company C, First Minnesota Volunteer Heavy Artillery in October 1864. The morning Woolson left, he watched thirty-eight Lakota warriors hanged in Mankato.[3]

Thousands of Duluthians paid their last respects to local resident Albert Woolson on August 6, 1956. Woolson was the last veteran of the Union Army to die, ninety-one years after Robert E. Lee surrendered the remnants of the Army of Northern Virginia to Ulysses Grant at Appomattox Courthouse, Virginia. Duluth News-Tribune, NEMHC S2386 B21f17

For the remaining six months of war, Woolson's regiment was attached to the Army of the Cumberland. The regiment primarily saw garrison duty around Fort Oglethorpe, the northwest Georgia post that was a few miles from the killing ground at Chickamauga. Woolson, like his father, was a musician and served as a regimental drummer during his time in the Army, which lasted just less than a year.

Woolson returned to Minnesota and was a carpenter for much of his life. He later moved to Duluth to be with several daughters and died there of lung congestion on August 2, 1956.[4] Thousands turned out to pay their respects to Woolson on August 7, 1956. His services were held in the Duluth Armory, and the funeral cortege wound through crowds on every street corner in Duluth's East End on its way to Woolson's final resting place at Park Hill Cemetery.[5]

Woolson's passing was commented on by President Eisenhower, an old soldier himself. "The American people have lost the last personal link with the Union Army," the president said in a prepared statement. "His passing brings sorrow to the hearts of all of us who cherished the memory of the brave men on both sides of the War Between the States."[6]

he Ramon de Larrinaga *slips nderneath the Duluth Aerial Bridge n a foggy, cool Sunday morning in ay 1959 to open the Twin Ports to orld commerce via the St. Lawrence eaway.* NEMHC S2386 B1f9

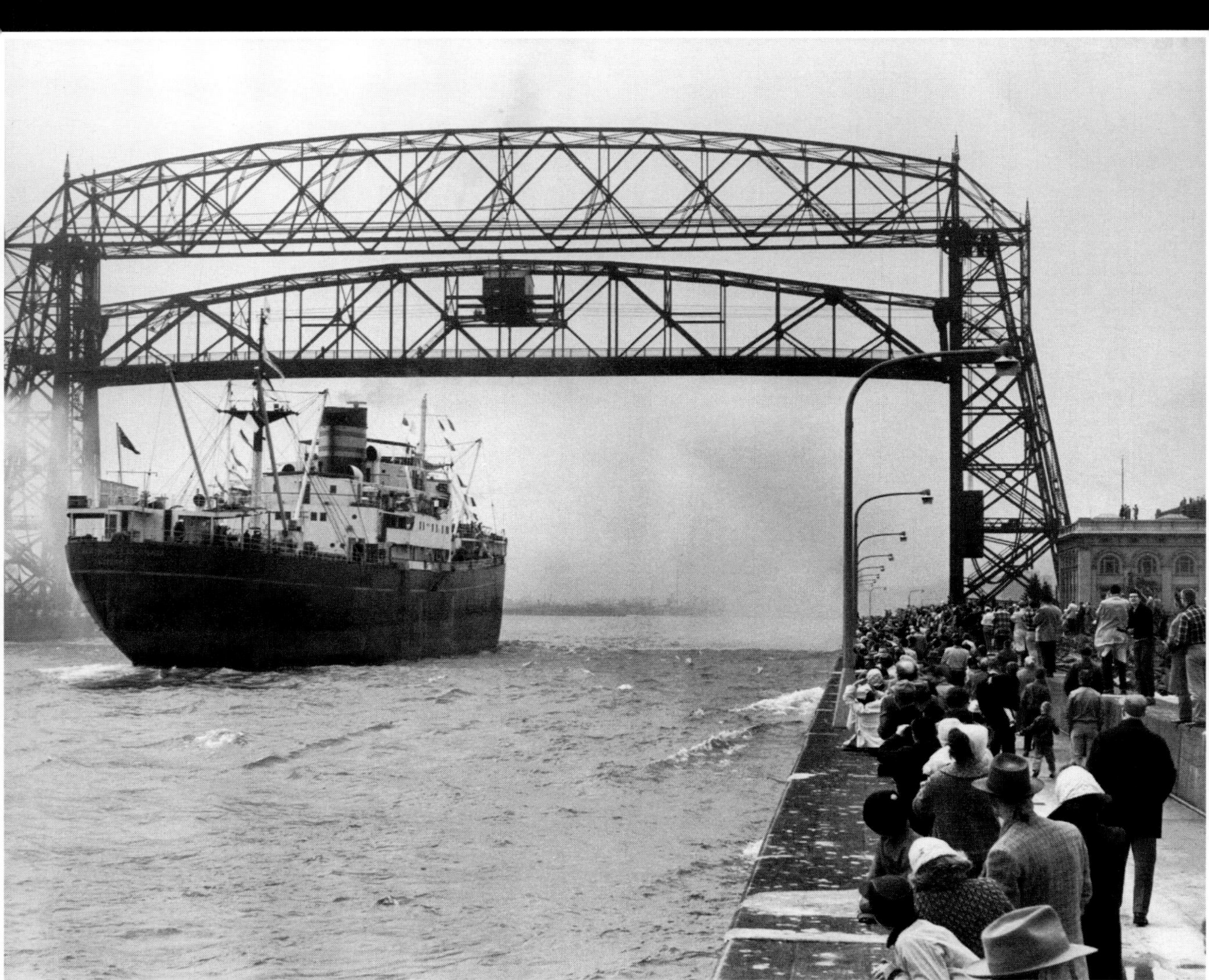

Secretary of the Army Wilbur M. Brucker headed a delegation of political and military dignitaries at the funeral. In presenting a United States flag to Woolson's daugher, Gertrude Virginia "Trudy" Kobus, Brucker noted that "this is no ordinary flag nor ordinary day. This is the last flag that shrouded the remains of the last veteran of the Union Army."[7]

The Beginning of an Era

Albert Woolson's death closed the curtain on an era that almost seemed shrouded in the mists of time. But by the summer of 1956, Duluth, the Iron Range and St. Louis County were eagerly awaiting the opening of a new era that had the potential of ushering in a wave of prosperity for the natural resources economy of the Upper Great Lakes region.

Congressional approval of the Wiley-Dondero Act in 1954 committed the United States to the construction of a bi-national seaway from the Head of Lake Superior to the Atlantic Ocean five years hence. News that the St. Lawrence Seaway would make Duluth-Superior an international port created a groundswell of enthusiasm in St. Louis County and the state. The foundation was quickly laid for the construction of harbor facilities able to handle the expected increased volume of waterborne cargo.

Arthur M. Clure, an admiralty attorney and expert in the sometimes convoluted world of St. Louis County mineral real estate, headed a new Seaway Port Authority of Duluth established in late 1954 by the Duluth City Council. In 1956, the Port Authority hired its first port director. Robert T. Smith, formerly the president of a New Orleans stevedoring firm, immediately began the complicated task of planning the construction of a major marine terminal on the Duluth waterfront. Estimated to cost $10 million, the new terminal would be financed with support from the state, the county and the city of Duluth.[8]

Clure, whose name would grace the new terminal, died unexpectedly in the spring of 1956. By that time, the wheels were set in motion to acquire the land and build the Clure Public Marine Terminal. Governor Orville Freeman visited the Twin Ports in December and promised that he would work overtime to convince the Minnesota Legislature to appropriate at least half the money for the maritime facilities that would make Duluth an international port.[9]

Some thirty months after Freeman gave his assurance that the state would do its part to ensure the successful completion of the new marine terminal, the M.V. *Ramon de Larrinaga* sailed under the Duluth Aerial Bridge on the misty morning of May 3, 1959.[10] The first upbound ship to cross the newly opened Seaway from Montreal to the Head of the Lakes, the *de Larrinaga* commenced a new era that exposed St. Louis County to the world.[11]

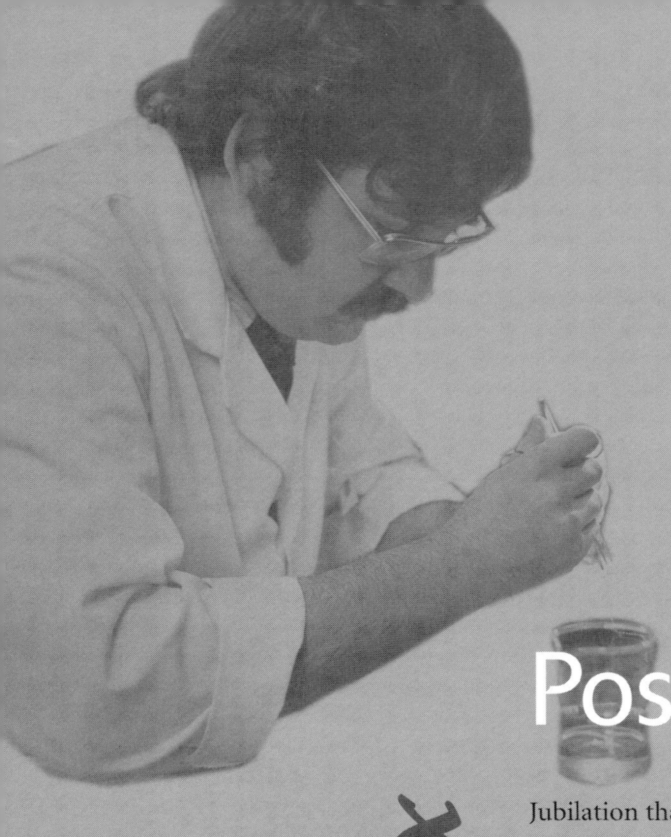

chapter six

Postwar Challenges

Jubilation that the United States had triumphed over Imperial Japan and Nazi Germany in World War II was tempered by three sobering concerns in the fall of 1945. First and foremost was the fear that America would soon be at war with the Soviet Union, its former ally and only other world-superpower. A second fear was that the American economy wouldn't be able to absorb the millions of soldiers, sailors and Marines returning from the war fronts around the world.

The first two concerns were national in scope, and as it turned out, less earth-shaking than many had originally feared. Josef Stalin's Soviet Union was more interested in creating buffer states in Eastern Europe than in challenging American interests in Western Europe. When confronted by resolute U.S. action during the Berlin Airlift, for example, the Soviets backed down.

And far from reverting back to Depression-era economic conditions, the American economy took off like a rocket in the years immediately following the war. Millions of veterans took advantage of the GI Bill of Rights to acquire an education, and more millions used their veterans' benefits to buy homes. Since there had been relatively little to spend money on during the past fifteen years, the economy was awash in cash. Defense factories made a rapid transition to producing automobiles, appliances, and the other consumer durables that Americans had been doing without for a decade-and-a-half.

All the production for the consumer economy required steel, and steel required iron ore. That led to the third sobering postwar concern, especially for those living and working in St. Louis County. Geologists in 1945 suspected that the United States was about to run short on iron ore.

Jubilation that the United States had triumphed over Imperial Japan and Nazi Germany in World War II was…

In 1945, the big fear on Minnesota's Mesabi Iron Range was that the high-grade natural ore that had gone down the Great Lakes from such mines as the Mountain Iron during World War II would soon be exhausted. NEMHC S3021 B16f49

Fortune magazine raised the issue in its November 1945 edition. In the six years since 1940, almost a half-billion tons of high-grade ore had gone down the Lakes from the Lake Superior ore ranges, including the Vermilion and Mesabi, *Fortune* reported.[1] The magazine article went on to quote geologists, who predicted that at the current rates of depletion, the Mesabi might well run dry of rich natural ores in as little as a decade.

There was evidence that the steel industry was prepared to look elsewhere for its supplies in the postwar era. U.S. Steel and Bethlehem, the two largest steel companies in the United States, chose to build new blast furnaces at Fairless Hills, Pennsylvania, and Sparrows Point, Maryland, in the early 1950s, instead of at mills on the Lower Great Lakes. Observers of the industry considered it a tacit admission that the two steel companies were considering getting their ore supplies from vast new deposits discovered during the war in Labrador, Brazil, and Venezuela.[2]

The flow of iron ore from the Mesabi and Vermilion was unabated during the immediate postwar years. An average of 50 million tons a year crossed the docks in Duluth-Superior each year between 1945 and 1950, most of it destined for cars, refrigerators, and reinforcing bar for the surging American economy.[3]

The North Korean invasion of South Korea in late June 1950 unleashed the floodgates of Lake Superior ore. By 1953, the Mesabi was shipping an all-time record of more than 60 million tons of iron ore to the mills on the Lower Lakes for America's participation in the Korean War.

Even the staunchest supporters of the Mesabi Range had to wonder if *Fortune* hadn't been right back in 1945. Since the article appeared eight years before, another 500 million tons of Mesabi and Vermilion Range ore had been shipped. Fortunately, the University of Minnesota and the steel industry were ready to unveil a new technology that would transform the Lake Superior iron ranges.

Mesabi Range producers continued to put new properties into production during the postwar years, such as the Fayal Annex in 1947. But at the same time, domestic steel customers were actively searching for new, rich ore deposits in South America and Africa. NEMHC S3021 B16f21

The Friday night fish fry became part of the social fabric in the small towns of rural St. Louis County following World War II. NEMHC S2386 B12f28

Loaded ore cars at the Proctor yards of the Duluth, Missabe and Iron Range Railway during the 1940s are grouped according to the chemical analysis of the ore they contain before they are shipped to the docks in West Duluth. Rykken, NEMHC S4525 "O"

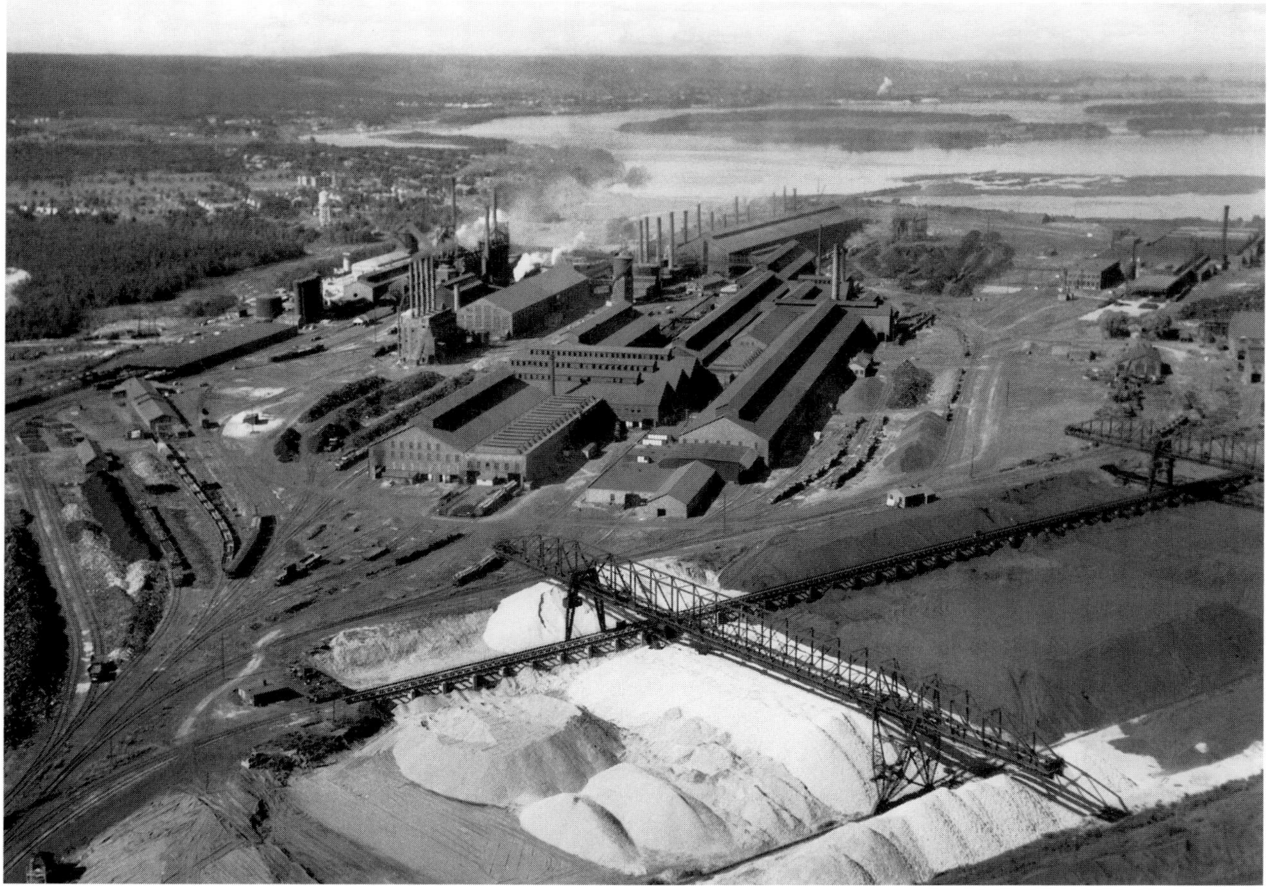

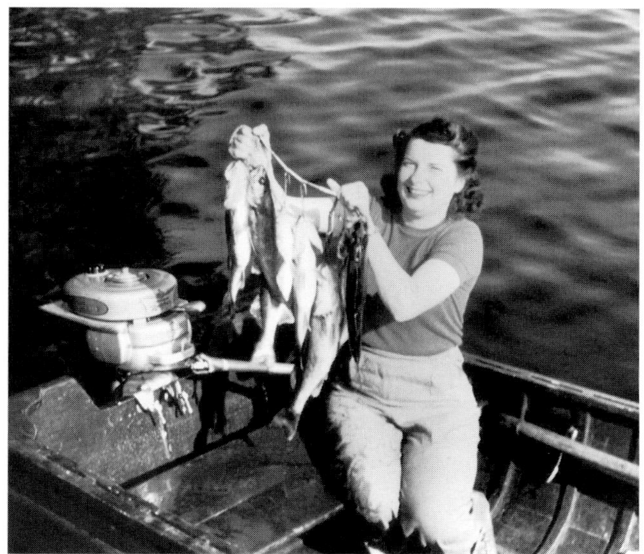

In 1950, U.S. Steel Corporation's Minnesota Steel Plant in Morgan Park was part of an industrial complex that made Duluth and St. Louis County an integral part of the nation's economic strength. The region would soon be called upon to repeat its World War II performance and provide iron, steel, and manufactured goods for U.S. participation in the Korean Conflict. NEMHC S2386 B30f7

The completion of all-weather roads, many of them built initially to allow local loggers access to federal and state timber sales, opened up much of the central and northern parts of the county to tourism in the years followed World War II. News soon reached residents in the Twin Cities, Milwaukee, Chicago, and Indianapolis that Minnesota's Arrowhead boasted some of the best fishing in the Midwest. NEMHC S2386 B35f6

Technology and Timber Harvesting

Technology also was transforming another staple of the St. Louis County economy in the years after the war. Timber harvesting underwent a major shift during the period from the 1940s to the 1960s.

The camp system that had prevailed since the 1910s disappeared rapidly in the postwar era, primarily because all-weather roads made stands of timber available to local teams of loggers utilizing pickup trucks and bulldozers, both of which had been developed during the war. Another technology introduced on the fighting fronts—the chainsaw—allowed two or three men to do the work of a dozen before the war. The chainsaw revolutionized timber harvesting.

The development of a pulp and paper industry in the counties adjacent to St. Louis County during the 1950s—at Grand

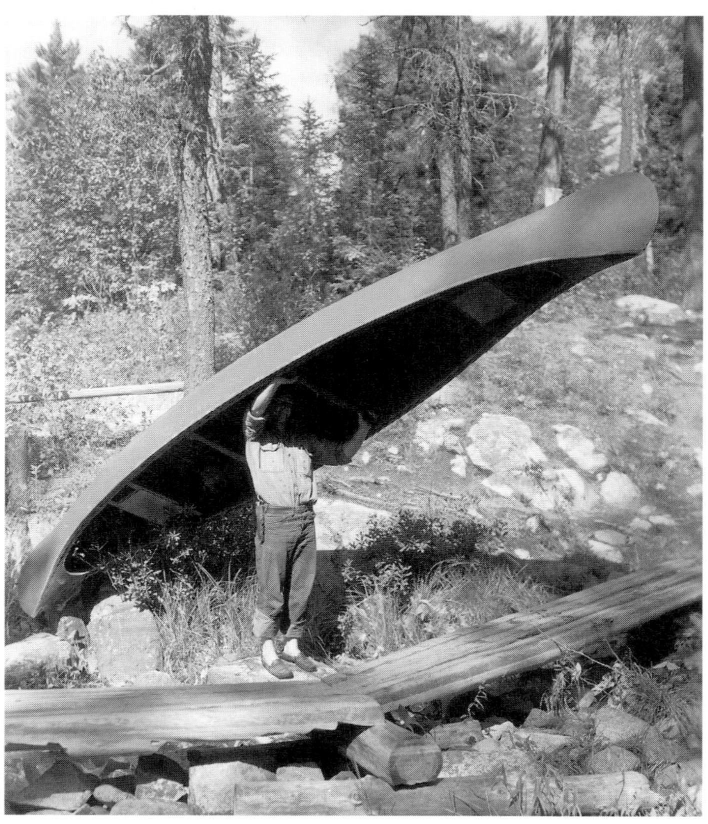

Camping in the Boundary Waters Canoe Area Wilderness became an increasingly popular summer pastime during the 1950s and 1960s. NEMHC S2386 B34f20

The development of riveted aluminum canoes—an outgrowth of aircraft fuselage construction during World War II—made the watercraft affordable for a generation of Americans who thought portaged the lightweight canoes past rapids and deadfalls in the Boundary Waters. M. J. Humphrey Photo, NEMHC S2386 B34f20

Rapids and International Falls—created a new model for logging in St. Louis County and northeastern Minnesota. By the 1960s, new rubber-tired logging equipment—skidders and feller-bunchers—were allowing logging companies in such rural St. Louis County communities at Gheen, Buyck, Biwabik, and Meadowlands to do the work of an entire logging camp twenty years before.

Creation of a Wilderness

The mechanization of the industry was brought about, at least partially, by a labor shortage that grew worse each year after 1960. Logging also was changed in other ways. By the 1960s,

By 1963, Minnesota boasted some of the most powerful and well-respected Democratic politicians in America. Here, Senator Eugene McCarthy, former Governor and Secretary of Agriculture Orville Freeman, and Senator Hubert Humphrey share the podium with President John Kennedy in September 1963, just two months before the president was assassinated in Dallas. Humphrey would serve as President Lyndon B. Johnson's vice president and narrowly lose the presidency to Richard M. Nixon in 1968. McCarthy would knock LBJ out of the race in the 1968 presidential primaries. Freeman would be responsible for helping create the Boundary Waters Canoe Area Wilderness.
NEMHC S2386 B21f7a

much of northern St. Louis County was in the process of being transformed into the Boundary Waters Canoe Area Wilderness (BWCAW).

Conservation of the county's abundant natural resources was rooted in the past. A huge stretch of the U.S.-Canadian border in Minnesota from just east of International Falls almost all the way to Lake Superior was little changed from the era of the fur trade two centuries before. Much of the land had not been stripped for mining, mainly because the Vermilion Range and its high-grade ore petered out northeast of Ely. More accessible stands of timber farther south in St. Louis County meant that much of the area retained its tree cover from at least the late nineteenth and early twentieth centuries.

The land that would become the BWCAW had been under federal protection since 1909 when the federal government established the Superior National Forest in a broad swath across the northern end of St. Louis County. A 1926 wilderness policy allowed for continued timber harvesting in the national forest, but passage of the Shipstead-Newton-Nolan Act four years later restricted logging adjacent to the forest's watersheds.[4]

The U.S. Environmental Protection Agency's Water Quality Laboratory on Lake Superior just northeast of the Duluth city limits is one of the nation's major research facilities for the study of environmental toxicology and chemistry.
NEMHC S3065 B1f23

Part of the national forest was designated the Superior Roadless Area in 1938, and in 1941, the northern half of what would become the wilderness area was protected from all timber harvesting. Airplane flights over the area were banned in 1949. Senator Hubert Humphrey sponsored legislation in 1956 that would culminate in the area's designation as the Boundary Waters Canoe Area Wilderness eight years later.[5]

When President Lyndon B. Johnson's signature made the Wilderness Act law in 1964, the new legislation had an immense impact on St. Louis County. Suddenly, more than 360,000 acres across the northern fifth of the county were forever off-limits to logging, one of the county's economic

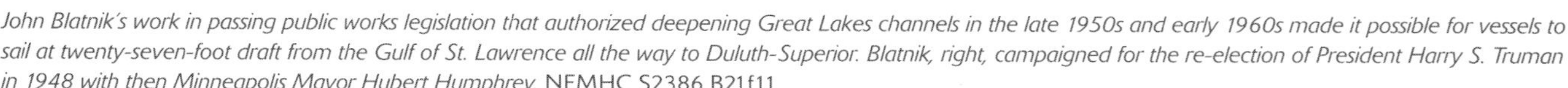

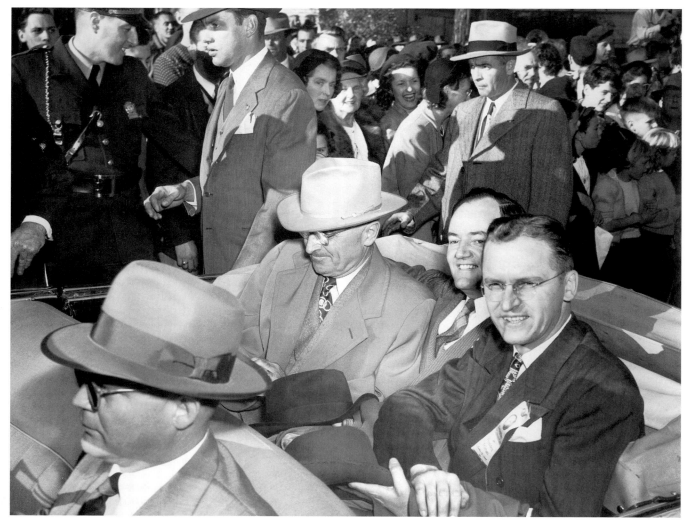

John Blatnik's work in passing public works legislation that authorized deepening Great Lakes channels in the late 1950s and early 1960s made it possible for vessels to sail at twenty-seven-foot draft from the Gulf of St. Lawrence all the way to Duluth-Superior. Blatnik, right, campaigned for the re-election of President Harry S. Truman in 1948 with then Minneapolis Mayor Hubert Humphrey. NEMHC S2386 B21f11

Left: Dave Durenberger was part of the Independent-Republican tandem that held Minnesota's two U.S. Senate seats for most of the 1980s. NEMHC S3065 B2f23

Congressman James L. Oberstar of Chisholm succeeded his boss, John Blatnik, as Eighth District congressman in 1974. Blatnik's administrative assistant, Oberstar spoke fluent French and was an expert in the problems of Haiti. As St. Louis County celebrates its 150th anniversary in 2006, Oberstar and Blatnik have represented the Eighth District for sixty years straight, 40 percent of the county's history. NEMHC S3065 B2f26

The opening of the St. Lawrence Seaway in 1959 led to the re-establishment of the Seaway Port Authority of Duluth. The Port Authority's first order of business was construction of a public marine terminal, which opened in 1959 and was named in honor of the port's first chairman, Arthur M. Clure, a Duluth admiralty attorney.
NEMHC S3065 B1f50

mainstays. The Boundary Waters would soon prove to be an economic boon for the county's growing tourism industry. The next year, Secretary of Agriculture Orville Freeman, a former governor of Minnesota, nearly doubled the no-cut zone in the BWCAW to 667,000 acres.[6]

The fight over what simply became known as the Boundary Waters pitted rural residents of the northern quadrant of the state against urban folk in the southern half of the state. Loggers wanted their livelihood maintained; Twin Cities residents wanted an unspoiled wilderness close to home. There were really no right or wrong sides to the issue, and in the end both sides compromised for the mutual benefit of the state.

The Kabetogema Peninsula just west of St. Louis County soon was designated the centerpiece for what would become Voyageur's National Park. By the late 1960s, more than 2 million acres of land stretching from Rainy Lake two hundred miles east to Lake Superior would come under federal ownership and control as either national park or wilderness area.[7]

John Blatnik's Influence

The point person for county unhappiness over federal land policies was a decorated war veteran who dominated congressional politics in the area for nearly a quarter-century. John Blatnik was an Iron Ranger who parachuted into war-torn Yugoslavia with the Office of Strategic Services (OSS) and came back to Minnesota to claim the Eighth District seat in Congress on his first try in 1946.

The son of Slovenian immigrants, his ability to speak fluent Slovene put him in good stead when he was serving as an OSS liaison to partisans fighting for Marshal Tito against the Germans in 1942 and 1943. The first American of Slovenian descent elected to Congress, Blatnik wielded enormous influence in northeastern Minnesota. Elected as a Democratic-Farmer Labor (DFL) candidate, Blatnik was early-on an ally of Hubert Humphrey, who was mayor of Minneapolis when Blatnik made his first successful run at Congress in 1946.[8]

Blatnik grew up poor on the Iron Range, and he knew what it was like to live through a Great Depression that shut down most of the region's iron mines. From the first, he was a champion of labor and a supporter of the civil rights movement, providing the model for DFL politicians in northeastern Minnesota for the next sixty years.[9] Blatnik also was a strong advocate of environmental legislation. As early as 1948, he began sponsoring water quality legislation in Congress, legislation that wasn't passed until his final terms in office in the early 1970s.[10]

Blatnik, however, did not let environmental legislation stop development that would benefit his St. Louis County constituents. Blatnik always tried to mediate disputes between wilderness legislation and logging interests.

In the 1960s and 1970s, he was perhaps the staunchest defender of Reserve Mining Company, which at the time was at the center of a huge environmental fight over its disposal of taconite tailings into Lake Superior. Perhaps the biggest irony of the whole affair was that scientists from the federal water quality laboratory in Duluth—a plum government project that Blatnik brought home to the district—were responsible for providing the evidence that caused the courts to order Reserve to stop dumping the tailings into the Lake.[11]

John Blatnik rose through the seniority system of the U.S. House of Representatives to chair the influential House Public Works Committee. As an Iron Ranger and a former education advisor, he fully understood the importance of public works for the Eighth Congressional District.[12] During his twenty-seven years in Congress, Blatnik brought dozens of federal projects to his District.

Duluth was an early recipient of interstate highway funding. One of the first interstates completed in Minnesota was I-35 from the Twin Cities north to the Head of the Lakes. It was on Blatnik's watch that U.S. Highway 53 from Duluth to the Mesabi Range and north to International Falls, as well as U.S. Highway 169 across the Range, were improved, widened, and four-laned. Interstate 535, which connected Duluth and Superior, was anchored by a $21 million high bridge over the harbor; 90 percent of the cost was paid by the federal government. The 1961 completion of the new bridge, which officially would be named for Blatnik ten years later, was the initial link in a federal system that would tie St. Louis County to Minnesota and the rest of the nation with a ribbon of concrete.[13]

Linking St. Louis County to the World

Blatnik's greatest public works project was the St. Lawrence Seaway. Several residents of the county were instrumental in keeping alive the dream of connecting the Great Lakes with the Atlantic Ocean. As early as the World War I years, the Great Lakes–St. Lawrence Tidewater Association began lobbying Congress for the deep-draft waterway that would open the heart of North America to global maritime commerce. Organized by Charles P. Craig, a Duluth attorney, the Tidewater Association was a tireless advocate of building the bi-national Seaway.[14]

Clarence LaLiberte, later the longtime president of Duluth's Cutler-Magner docks, made dozens of trips to Washington, D.C., on behalf of the Tidewater Association. The project proceeded in fits and starts. In 1932, the United States and

Duluth's cargo base was diversified during the early 1960s, much of it from Lower Lake ports and not only because of the completion of the St. Lawrence Seaway. At that time, Canal Park was still dominated by light industrial activity that wouldn't be reclaimed for tourism purposes until the 1980s and 1990s. Florman, NEMHC S2386 B15f14

The opening of the St. Lawrence Seaway made Duluth an international seaport. Bagged cargo ranging from coffee to spices was unloaded by local longshoremen at the Port of Duluth during the 1960s. Stevedores, pilots, linehandlers, customs agents, grain inspectors, and dozens of other port-related professionals earned their living in Duluth and Superior. NEMHC S3065 B1f50

The Volkswagen Beetle—Germany's contribution to automotive progress—made its initial appearance in the Twin Ports aboard a saltwater vessel sometime in the 1960s. By that time, trucking company executive Malcolm McLean was perfecting the idea of shipping cargo in containers. Container traffic revolutionized world shipping; unfortunately, the trend took hold on the coasts and skipped the Great Lakes. NEMHC S3065 B1f50

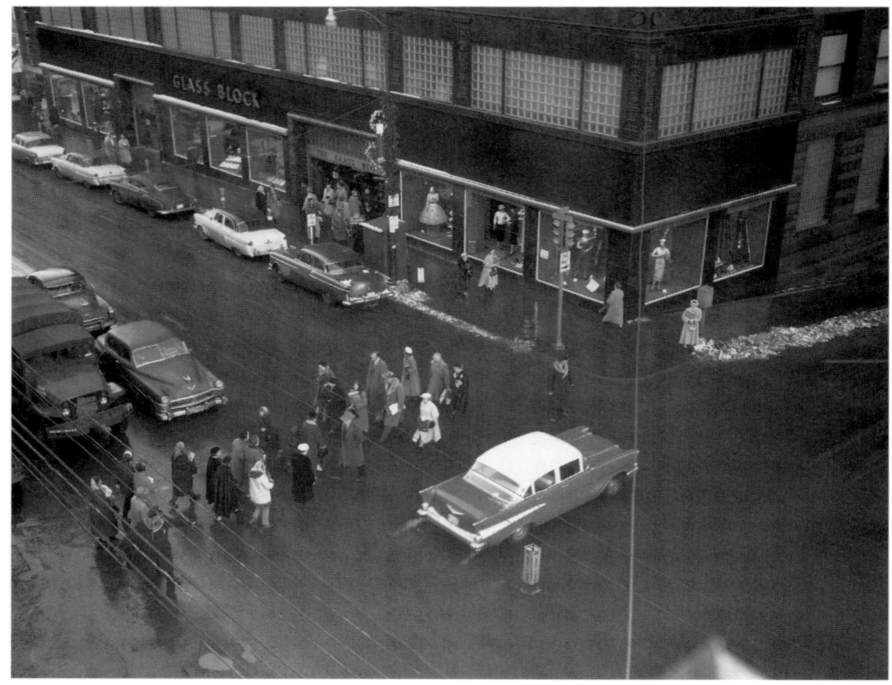

Duluth's Glass Block Department Store in 1957. In the years after World War II, Duluth established itself as a shopping, health care, and secondary educational center for much of northeastern Minnesota and northwestern Wisconsin. NEMHC S3766 B2f7

Canada agreed to build a joint seaway, but there was no money in Depression-era budgets to actually accomplish the physical construction of the project. President Franklin D. Roosevelt resurrected the Seaway as a defense measure in 1941, but Midwestern and East Coast railroads—which FDR needed to win the war—strenuously opposed the measure.[15]

Ironically, postwar concerns about depletion of Mesabi Range natural ores finally helped make the Seaway a reality. In 1951, with American troops fighting for their lives in North Korea's Yalu River Valley, Canada informed the U.S. government that it was going to build a Seaway, whether the United States elected to participate or not. Part of the Canadian decision was based on the not unrealistic hope that the Seaway would be the spur to developing the rich iron ore resources of Quebec and Labrador.[16]

The Tidewater Association swung into action for one last attempt at convincing Congress to join the Canadians in building a Seaway. Julius Barnes, who had built ships in Duluth during World War II and had helped Herbert Hoover feed Europe in the aftermath of World War I, lent his name and checkbook to the fight. He was joined by Lewis G. Castle, a Duluth banker and the son-in-law of Alexander McDougall. Together, the two men put together a last-ditch lobbying and public relations offensive that helped convince Congress to pass the Wiley-Dondero Act in the spring of 1954.[17]

Five years later almost to the day, the M.V. *Ramon de Larrinaga* steamed through the Duluth Ship Canal on a misty May morning to complete its first-ever journey to the Head of

Richard Leslie Griggs, seen here in 1944, was a banker, real estate investor, and one-time owner of the Poker Fleet. He was also the man who donated the land for a new campus for the University of Minnesota–Duluth at the end of World War II. NEMHC Portrait File

Top right: Matti Kaups, shown here teaching a geography class at UMD, was emblematic of the Finnish heritage of St. Louis County. Kaups was an outstanding teacher and respected scholar, with several books on Finnish settlement of the United States to his credit. NEMHC S3065 B2f4

Top left: The State Normal School, seen here in 1906, opened in 1895 and evolved into the Duluth State Teachers College in 1921, the University of Minnesota, Duluth Branch in 1947, and UMD in 1959. The prominent Old Main Building, pictured at left, burned in an arson fire the morning of February 23, 1993. NEMHC S2386 B9f5a

Bottom: Thanks to support from the administration of Governor Rudy Perpich and the northeastern Minnesota delegation in the Minnesota Legislature, the University of Minnesota Duluth matured into a well respected institution of higher education that offered graduate programs in engineering and medicine. The University's Division 1 hockey program, under Coach Mike Sertich, made it to the finals of the NCAA hockey tournament and routinely beat the Minnesota Gophers, its bigger cousin in Minneapolis. By the 1990s, UMD was attracting growing numbers of students from the Twin Cities suburbs. NEMHC S3065 B2f8

the Lakes along the twenty-two-hundred-mile-long Seaway. *National Geographic* called it "one of the most incredible engineering and construction jobs men have ever attempted, and in some ways the hardest."[18]

With a price tag of nearly a half-billion dollars, the Seaway was one of the most expensive construction projects ever accomplished on the North American continent. For more than four years, as many as twenty-two thousand U.S. and

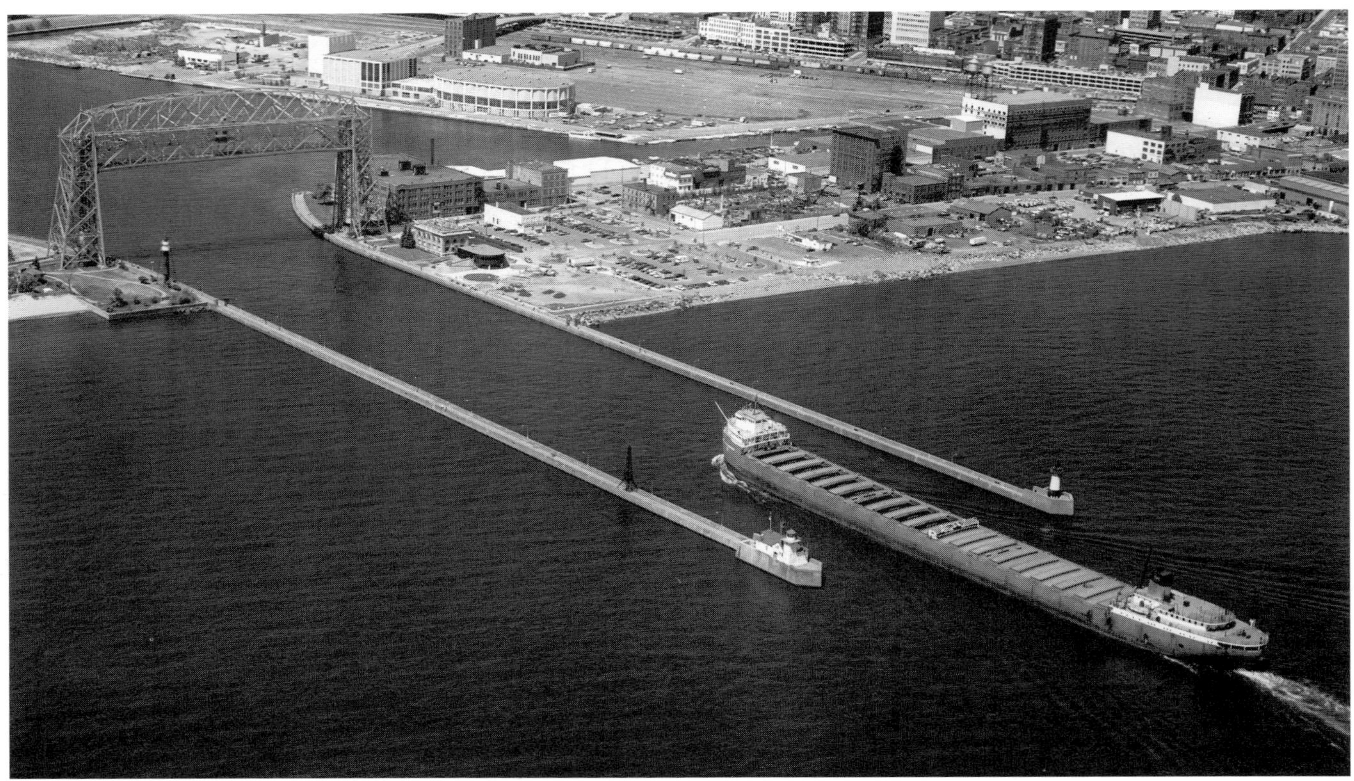

The 1960s and 1970s were prosperous years for the St. Louis County seat, Duluth. A straight-decker approaching the Aerial Lift Bridge in June 1973 would pass by the brand-new Duluth Arena-Auditorium as it slipped beneath the bridge. The construction of the Arena-Auditorium was the first step in a quarter-century effort to rehabilitate the city's waterfront. Canal Park on the Duluth side of the Lift Bridge in 1973 was still a mish-mash of seedy bars, flophouses, scrap yards, and industrial property. Ralph Knowlton, NEMHC S2386 B15f15

Pioneer Hall, site of one of the premier curling rinks in the Midwest, is shown here under construction in Duluth in 1976, ten years after the completion of the adjacent Arena-Auditorium complex. NEMHC S2386 B1f4

Davis' research led to the development of technology to produce taconite pellets, which became the feedstock for most of North America's steel industry blast furnaces during the 1960s and 1970s. Knowlton, NEMHC S2386 B15f15

Canadian construction workers moved more than 210 million cubic yards of rock and dirt.[19]

John Blatnik hovered over the project like a mother hen. His 1957 legislation authorizing the U.S. Army Corps of Engineers to spend nearly $150 million deepening 150 miles of connecting channels on the Great Lakes from twenty-five to twenty-seven feet, as well as iron ore harbors such as Duluth-Superior and Two Harbors, ensured that St. Louis County and its iron ore economy would be major beneficiaries of the Seaway.[20] When the Corps deepening project was completed in 1964, Duluth-Superior had a brand new harbor capable of handling the largest Seaway ships.[21]

A New Era

The opening of the Seaway ushered in a new era for Duluth and St. Louis County. The port city became one of the nation's major grain ports. By the 1960s, it was common to see sailors from a host of nations shopping in downtown Duluth. When relations with the Soviet Union thawed during the administration of President Richard M. Nixon in the 1970s, Duluth was one of the primary ports Soviet ships involved in the Russian grain trade called upon. The Soviet grain terminal ports on the Black Sea had many of the same depth restrictions as the St. Lawrence Seaway.

With the opening of the Seaway, Duluth increasingly became a center for shopping, education and health care, not only for St. Louis County, but for the tri-state area of northeastern Minnesota, northwestern Wisconsin, and the western tip of the Upper Peninsula of Michigan. The city reached its peak population in 1960 of nearly 107,000 people. Although Duluth lost population during the next twenty years as the area's economy was restructured, demographics created a demand for housing in the rural townships surrounding the city. The population of Duluth and the townships in 1980 was slightly larger than the population of Duluth and the townships in 1960.[22]

One factor driving the port city's emergence as a regional center was education. The University of Minnesota Duluth evolved after the war from what had been Duluth Teachers' College into a major component of the University of Minnesota. In 1964, the university reported an enrollment of thirty-two hundred students and was in the midst of a

E. W. Davis devoted forty-five years to the development of Minnesota's low-grade taconite deposits. Davis retired in 1955 as the head of the University of Minnesota's Mines Experiment Station. NEMHC S2386 B1f4

The first full-scale taconite plant in Minnesota was Erie Mining Company, which began producing pellets in the mid-1950s. The mine and pelletizing facility were located at Hoyt Lakes, on the far eastern end of the Iron Range. Instead of shipping pellets to Duluth or Superior, Erie shipped its pellets to Taconite Harbor on the North Shore of Lake Superior, midway between the Head of the Lakes and the Canadian border. NEMHC S2386 B30f3aa

ten-year, $17.7 million building program. At the time, the College of St. Scholastica had a student body of five hundred women. By the time St. Scholastica went coeducational and admitted its first fifty male students in 1970, the college student population in Duluth was pushing five thousand.[23]

The Gateway Urban Renewal Project, completed with federal and state money during the 1960s, resulted in a reconstruction of most of the downtown business district. Dilapidated hotels and boarding houses along West Michigan Street were demolished to make way for parking for downtown office workers. The construction of the $2.5 million National Water Quality Laboratory on the North Shore near Lester River brought high-paying federal jobs to the city, and the 1966 completion of the $6.1 million Duluth Arena-Auditorium capitalized on the Head of the Lakes' growing tourism business.[24]

St. Luke's Hospital and St. Mary's Hospital, the two primary healthcare facilities in Duluth, created the infrastructure for a major regional medical center in the Twin Ports. St. Luke's built an east wing in 1950 and added a west wing in 1969.[25] St. Mary's constructed a new, nine-story west wing in 1957, razed the old east wing in 1967 and replaced it with a new, $5.5 million, seven-story tower in 1969.[26]

Together, the two hospitals spent more than $11 million in the late 1960s alone to modernize their facilities and bring bed capacity on the two campuses to nearly 900 beds.[27] With the nearby Duluth Clinic and Polinsky Rehabilitation Hospital, St. Mary's and St. Luke's could truly lay claim by 1970 to offering the finest health care services north of the Twin Cities.

Reserve Mining Company was another early taconite operator. Reserve, which extracted ore from the Peter Mitchell Pit near Babbitt well east of most Mesabi Range natural producers, elected to ship the ore downhill to Silver Bay on the North Shore for processing, palletizing, and shipping. Beginning in the 1950s, Reserve released millions of tons of taconite tailings into Lake Superior, precipitating one of the first major battles by the nation's environmental movement. U.S. Judge Miles Lord eventually ruled Reserve could not continue dumping taconite tailings into the Lake. In the early 1980s, the taconite company began on-land disposal of its tailings at Mile Post 7. NEMHC S2386 B30f3aa

The Taconite Revolution

The prosperity that permeated the Twin Ports in the 1950s and 1960s was shared in the mining regions of St. Louis County.

In the early 1950s, there were still underground workings on the Mesabi and Vermilion Ranges, such as this small mine at Eveleth, but underground operations were becoming a thing of the past in St. Louis County. NEMHC S2386 B16f22

Boosters pulled out all the stops to pass the Taconite Amendment during the 1964 elections. Supporters of the measure urged voters to indicate their backing by using taconite stamps on their mailings. As it was, the measure passed handily, with 80 percent of the vote. NEMHC S4572,291

Postwar fears that the Mesabi and Vermilion Ranges would become strewn with ghost towns when the rich natural ore inevitably ran out proved to be unfounded. The concerns about the natural ore being depleted were true. But natural ore was replaced by a low-grade ore that was concentrated, rolled into marble-sized pellets and fired in giant kilns on the Mesabi Range. The resulting taconite pellets revolutionized steel making and provided the Iron Ranges with a second wind, some sixty years after they had first begun commercial production.

Taconite was a low-grade iron formation that earlier miners in the region had discarded as waste rock. With an iron content of about 30–35 percent, taconite had no commercial value when hematite with an iron content approaching 75 percent was readily available.

As far back as the 1930s, Professor E. W. Davis had begun investigating the feasibility of concentrating taconite at his laboratory in the Mines Experiment Station at the University of Minnesota. By the 1940s, Davis had demonstrated that

Man Made Mountains of Minnesota

The Great Pyramid of Gizeh covers an area of thirteen acres (5 city blocks). Hibbing's Hull-Rust-Mahoning open pit iron ore mine has an area of 1535 acres. Approaching a BILLION TONS of material removed, 443 million gross tons of waste material has had to be piled high to almost overshadow our world-famous Mesaba Range.

The "Mountains" are being created by converted underground mines, giant electric and diesel shovels, fabulous trucks, endless belts, draglines, and locomotives of Paul Bunyan magnitude. With coordinated "ant hill" precision they expose the deeper and deeper bodies of High and Beneficiation grade ores at an accelerated annual rate. Visualize the challenge to supply our nation's demands met by our basic industry thru technological skill. Magnificance personified. Minnesota—Land of Paul Bunyan and Big Opportunity.

The Chamber of Commerce is cognizant of the evolution taking place in Iron Ore Mining. Beneficiation of lower grade ores currently produces one third of all the iron ore shipped. Foregin High Grade Ores and Taconite are competitive.

YOUR

Chamber of Commerce

506 E. Howard Street — Dial AM 3-7611

The Hibbing Chamber of Commerce called Hibbing "the Iron Ore Capital" in this 1952 pamphlet. NEMHC Pamphlet 295

One of the first taconite companies to begin commercial production following the passage of the Taconite Amendment was Eveleth Taconite. The company shipped its first pellets in November 1965, less than a year after the amendment was ratified. Eveleth Taconite was partially owned by Ford Motor Company, which processed taconite from the mine at its Rouge Steel Mills in Detroit. NEMHC S2386 B30f3aa

taconite could be ground to the consistency of talcum powder, which made the individual particles of iron in the rock susceptible to separation electromagnetically. Binding the iron particles with bentonite, a Wyoming clay, and rolling them in a pelletizer mill created iron pellets of 65–70 percent iron.[28]

Pickands-Mather, one of the older mining companies on the Minnesota Iron Ranges, put the first experimental taconite plant into operation in 1948 when it formed Erie Mining Company. Two years later, in 1950, U.S. Steel built its Pilotac plant at Virginia. For the next five years, Erie Mining Company, Pilotac, and Reserve Mining Company, a partnership that included Republic Steel and Armco Steel, struggled with the problem of concentrating the ground taconite. The goal of the researchers was to put taconite into a form that would be both easily transportable and suitable for use as feedstock in the blast furnaces at the lower end of the Great Lakes.[29]

The solution came with the discovery in the early 1950s that taconite powder could be mixed with a binding clay in a ball mill—not unlike a giant cement mixer—and then fired in a kiln. The resulting pellets could be transported in rail cars to the ore docks where they were gravity-fed into the pockets for loading ships. Unlike natural ore, taconite had the lion's share of its moisture baked out in the kilns. As a consequence, the pellets could be shipped year-round or stored on the ground without freezing into huge lumps.

Best of all, pellets were far superior to natural ore in the blast furnaces of the nation's steel industry. Dr. Peter Kakela, whose grandfather mined iron ore near Eveleth after arriving from Finland in 1898, explained that "taconite pellets simply work better in blast furnaces than the soft, red ores."[30] Kakela, a professor of resource development at Michigan State University and one of the nation's leading experts on iron mining, added that "pellets are the preferred form of iron ore because they improve productivity, conserve energy and allow greater quality control of iron production from the blast furnace."[31]

That was a secret the industry didn't necessarily want St. Louis County or the state of Minnesota to know a great deal about. The state of Minnesota, in fact, had shifted to a production tax in 1941 to encourage the development of low-grade iron ore. The tax was very low, about half of what natural ore

Virginia and Hibbing, the two largest communities on the Mesabi Iron Range, operated municipally owned light, water, and steam heating plants to keep costs low for city residents. NEMHC S2386 B31Bf1a

Thanks to the Iron Range Resources (IRR), a state agency set up by Governor Harold Stassen in 1941, residents of the Iron Range enjoyed first-rate municipal services and paid relatively low property taxes. Here, new sidewalks are installed at the North Leonidas location sometime in the 1950s. NEMHC S3021 B16f3

producers paid. Once Reserve Mining Company and Erie Mining Company had proved the commercial viability of taconite pellets in the late 1950s, a number of other iron and steel entities began planning new taconite plants.

But first, they wanted guarantees that the Minnesota Legislature would not push through hefty raises in production or occupation taxes. Beginning in the early 1960s, the iron and steel companies started a coordinated campaign to place an amendment to the Minnesota Constitution on the ballot for the 1964 election. The Taconite Amendment, as it became known, essentially guaranteed that the state would maintain the tax on pellets at roughly the same rate for the next quarter-century.[32]

The carrot that the iron and steel firms dangled in front of the state and its voters was the promise of a massive private investment in St. Louis County and adjacent Itasca County. As many as four companies announced plans to build new taconite plants in the area if the Taconite Amendment passed. It did.

Politicians from John Blatnik to Governor Karl Rolvaag to Senator Hubert Humphrey lined up behind the proposal. So did the United Steelworkers of America, which represented most of the miners in Minnesota's iron mining industry. In late October 1963, David J. McDonald, USWA's international president, made a special visit to the Iron Range to urge voters to support the amendment.[33] In November 1964, Minnesota voters approved the constitutional change by a 5–1 margin.[34]

The effect on the region's economy was immediate and breathtaking. Dave Gardner, the longtime public relations manager for Cleveland Cliffs, Inc., joined predecessor company Pickands-Mather as a public relations assistant at Erie Mining Company in 1965. Gardner recalled that "pioneers Erie and Reserve Mining were soon joined by Minntac (which would grow to be the largest taconite mine in the country); National Steel's Butler and National Steel Pellet plants; Eveleth Taconite; and Minorca."[35]

Between 1965 and 1969, Minntac, Eveleth Taconite, Butler, and National Steel Pellet brought on more than 13 million tons of taconite capacity.[36] Butler and National Steel Pellet technically were located in neighboring Itasca County, but the new mills contributed to the economy of St. Louis County communities such as Hibbing and Virginia. Following the Taconite Amendment, both Reserve and Erie announced plans to expand production. By 1967, the two original taconite plants had added almost 10 million tons of additional capacity.[37]

Between 1964 and 1967, taconite production in Minnesota—75 percent of it in St. Louis County—nearly tripled to

The Edmund Fitzgerald *sailed into history on the evening of November 10, 1975. The loss of the "Fitz" and her crew of twenty-nine sailors on Lake Superior's Whitefish Bay was a sober reminder of the danger of crossing the big lake "when the Gales of November come early."* NEMHC S2386 B42f5

more than 35 million tons. Another 10 million tons—some of it located at taconite plants in the Upper Peninsula of Michigan—came on line by 1969. Capacity in the Lake Superior taconite industry would almost double again during the 1970s.[38]

For nearly seventy-five years, iron ore mining in Minnesota had been seasonal employment. Mines and washing plants shut down when the weather turned cold and didn't reopen until spring. Miners learned to conserve their paychecks to tide them over the long winter.

But with the onset of taconite, the mining industry turned into a year-round, twenty-four-hour-a-day business. Taconite pelletizing plants ran best when they ran around the clock. And they employed huge work forces to blast, haul, and process the taconite at the giant mills. When Erie Mining Company expanded to 11 million tons in 1967, the company employed more than three thousand people in and around Hoyt Lakes, a company town that barely existed fifteen years before.[39]

Predictions that the steel industry in North America would be facing shortages of iron ore by the 1980s provided the spur to additional construction on the Mesabi Iron Range. In 1973, Pickands-Mather announced it was shutting the Mahoning Mine, a Mesabi producer since 1895. In its place, P-M and Bethlehem Steel told state officials they were building a $150 million taconite plant near Hibbing. The new plant, to be called Hibbing Taconite—and referred to as HibTac by just about everybody on the Iron Range—was scheduled to have a capacity of 5 million tons and to be in commercial operation by 1976.[40]

When Inland Steel's Minorca Mine came on line in 1979, the industry in Minnesota had a capacity of some 60 million tons.

More than fifteen thousand residents of St. Louis and adjacent counties derived a very good livelihood from the taconite mines and mills. Thousands more county residents earned a living in transporting the taconite by rail and lake vessel, or in the hundreds of subsidiary industries that served the mining industry. Thanks to Iron Range Resources (IRR), a government agency set up in 1941 to invest tax revenues in the Iron Range, St. Louis County residents paid no property taxes.

The opening of the Mesabi Range taconite industry during the decade siphoned off thousands of young county residents who might otherwise have gone into the woods for a living. Following the Tonkin Gulf Resolution in 1964, the Vietnam War played a key role in the labor shortage equation. More thousands of county residents joined the armed forces during the decade or gained a student deferment by attending college.[41]

Prosperous Times

Development of St. Louis County's taconite resources had a major impact upon the commercial shipping industry and Duluth, the county's largest city. Once the Taconite Amendment passed, the Great Lakes fleet—much of it headquartered in Duluth—realized that bigger and faster ships were necessary to move the anticipated shipments of pellets from the Head of Lake Superior and the North Shore downbound to the steel mills.

Part of the loss of jobs in the taconite industry in the 1980s stemmed from technological advances. Mine equipment was super-sized, with 240-ton trucks that ran on giant tires. Shovels tore overburden out of the earth at a 38-ton clip, and computers controlled the entire pelletizing process. By the late 1980s, the taconite industry required far fewer workers than it had just 10 years earlier. Jamar, NEMHC S3065 B2f7

In 1966, the U.S. Army Corps of Engineers began construction of a new lock at Sault Ste. Marie. The Poe Lock, named for a nineteenth-century Corps commander, was nearly twice the size of the existing locks at the Soo. The twelve-hundred-foot-long Poe Lock made possible a quantum leap in the size of vessels plying the Lakes when it opened to commercial traffic in 1969.[42]

The industry was quick to take advantage of the new lock. In the spring of 1972, Bethlehem Steel Corporation launched the *Stewart J. Cort,* the first of what would be known as the "thousand-footers." During the next seven years, twelve more thousand-footers joined the *Cort* in the Great Lakes fleet.

The new vessels revolutionized bulk commodity hauling on the Lakes. Able to haul upward of sixty thousand tons at a time, the thousand-footers displaced six of the earlier generation of six-hundred-foot vessels. Each thousand-footer was capable of making a round trip from Duluth to Gary, Indiana, discharging its cargo with self unloading booms aboard deck and returning to Duluth or Two Harbors in less than seven days.[43]

While the taconite industry on the Mesabi Range was adding workers by the thousands, the fleet—which had long been an employment option for many county residents—quickly shrunk in size. From 1971 to 1980, more than one hundred vessels of the two-hundred-vessel fleet were either consigned to scrap yards or long-term layup. Since each vessel eliminated from commercial operation was manned by a crew of approximately thirty-five, the layups and scrappings meant layoffs for more than three thousand Great Lakes sailors, many of whom were originally from Duluth or St. Louis County.[44]

Aside from the shrinking of the fleet, the 1970s were good years in Duluth and the Twin Ports. The Burlington Northern Railroad built a $40 million taconite loading dock in Superior, and the DM&IR upgraded its locomotive fleet and Proctor Yards. Cargill spent some $40 million to build a new, ultra-modern terminal elevator on Elevator Row in Duluth. Detroit

The wheels fell off the St. Louis County economy with frightening speed in the late 1970s and early 1980s. One of the early casualties of the restructuring in the steel industry was U.S. Steel Corporation's Minnesota Steel Plant in Morgan Park. By 1982, when the renamed USX Corporation had taken "steel" out of its name and transformed itself into a diversified energy and natural resources firm, Minnesota Steel was little more than a piece of America's industrial archaeology. Borne, NEMHC S2386 B30f8a

The domestic integrated steel industry was under assault during the 1980s from foreign steel producers and start-up minimills closer to home. The minimills used electric arc furnaces to melt scrap steel and make bar and rod products. Producers such as USX Corporation cast hot metal from blast furnaces into slabs and coils, shapes that the minimills couldn't make until 1987, when Nucor began producing thin slab steel at its Crawfordsville, Indiana minimill. Nimmo, NEMHC S2386 B30f8

The restructuring of the American iron and steel industry devastated the Great Lakes fleet. By the time this aerial photo of Howard's Pocket at the Superior foot of the Blatnik High Bridge was taken in the mid-1980s, dozens of lakes vessels were laid up in the Twin Ports, most never to sail again. The new thousand-footers which entered commercial service during the early to mid-1970s made vessels that had been built as recently as the 1940s obsolete. Between 1975 and 1990, the fleets scrapped half of the two hundred vessels that comprised the U.S. fleet. NEMHC S3065 B1f50

Edison and several partners built a state-of-the-art terminal in Superior to handle low-sulfur coal shipments from Montana through the Head of the Lakes.[45]

Even the tragic wreck of the *Edmund Fitzgerald* on Whitefish Bay of Lake Superior in November 1975 carried with it a cultural impact. Millions of Americans who would have been hard pressed to locate Lake Superior on a map knew all about Duluth and Superior when Gordon Lightfoot's sea chanty topped the music charts in the summer and fall of 1977.[46]

For a county that had had lived and died with extractive and natural resources industries for nearly 125 years, the 1970s were reminiscent of the boom times of the 1920s and the 1950s. Old-timers shook their heads and reminded anyone who would listen that iron and steel were the most cyclical of industries, and the good times couldn't possibly last forever.

Hitting Bottom

Nineteen seventy-nine was truly a miserable year for many Americans. Inflation in the late 1970s was running at double-digit rates, while interest rates spiked at about 15 percent. The administration of President Jimmy Carter struggled with strikes nationwide as workers demonstrated their displeasure with an economy that was eroding their paychecks.

One of those strikes took place in Duluth. A three-month walkout by Duluth-Superior grain millers crippled the Twin Ports' grain trade in the summer of 1979. The strike was settled in November, and things were just getting back to normal when the Soviet Union invaded Afghanistan. In retaliation, the Carter administration imposed a grain embargo on the USSR. It would be a decade before grain shipments through Duluth-Superior recovered.[47]

Jeno F. Paulucci was a tough Iron Range businessman who made a fortune selling canned Chinese food to American consumers. After selling his Chun King food line to R. J. Reynolds, Paulucci returned to Duluth and founded his Jeno's processed Italian food line. Paulucci later moved much of his food processing business to southeastern Ohio and got involved in Florida real estate development. NEMHC S3065 B2f26

In November 1979, U.S. Steel Corporation quietly announced a major cutback in operations. The country's largest steelmaker told business reporters it was closing ten old steel mills and shedding some thirteen thousand workers. It was just the beginning of a wrenching reengineering of the American iron and steel industry that would convulse St. Louis County for the next decade.

The restructuring of the industry took three years to develop momentum. Bethlehem, Armco, J&L Steel, LTV Steel, Republic, and other integrated steelmakers followed U.S. Steel's lead in the early 1980s, closing outdated mills and plants and laying off, in many cases permanently, thousands of steelworkers.

The reasons for the cutbacks were many and diverse. Steel companies had been complaining about unfair competition from low-priced imports since 1959, when the USWA shut down America's iron and steel industry for nearly four months in a landmark strike. Foreign steel flooded into the U.S. market to replace American steel idled by the lengthy labor problems.

Imports, however, were only part of the problem. Much of the steel arriving at U.S. ports was from Japan and Germany, former enemies whose steel mills had been blasted into oblivion during World War II. Postwar reconstruction of the devastated German and Japanese economies had resulted in the erection of ultramodern steel mills. As a result, German and Japanese steel often was competing against American steel that had been poured and rolled in mills built in the nineteenth century.

One pundit described the problem of aging mills succinctly when he noted that American steel would have been in far

Jack F. Rowe was a quiet, unassuming engineer who led Minnesota Power through the upheavals of the 1980s. Rowe helped the Duluth utility diversify into profitable water and telecommunications ventures, making the utility's common stock a darling of local investors. NEMHC S3065 B2f26

Jimmy Carter's selection of Walter "Fritz" Mondale as his running mate in the 1976 presidential election opened a new chapter in the political career of Iron Ranger Rudy Perpich. When Mondale, a former state attorney general and U.S. senator, was elected in November, his U.S. Senate seat fell vacant. Wendell Anderson, the state's DFL governor, resigned his office in early 1977, and Perpich, then the lieutenant governor, became governor. Perpich promptly appointed Anderson to Mondale's Senate seat, a move that so enraged Minnesota voters that they defeated both Anderson and Perpich in the 1978 elections. Anderson never ran for elective office in Minnesota again, but Perpich came back to claim the governor's office in 1982.
NEMHC S3065 B2f25

Rudy Perpich, seen here with Duluth Mayor Bob Beaudin about 1979, was known by his political enemies as Governor Goofy for his sometimes outrageous pronouncements and his economic development schemes. But Perpich convinced Alberta's Ghermezian Brothers to invest in what became the Mall of America in Bloomington, and during Perpich's two terms as governor of Minnesota was reported by many experts to have the best quality of life in the United States.
NEMHC S3065 B2f26

better shape in 1980 had Japan bombed Pittsburgh in 1941 instead of Pearl Harbor.[48]

Management in the American steel industry was top-heavy and bloated, and the USWA perhaps had been too successful, negotiating some of the richest contracts in American industry.

An even bigger threat than imported steel was lurking in the industry's backyard. In 1979, an upstart Charlotte, North Carolina company called Nucor Corporation was just beginning an expansion that would carry it to the top of the American steel industry. Led by Ken Iverson, a quiet engineer, the firm pioneered electric arc furnace technology that allowed Nucor to make steel from ferrous scrap rather than iron ore, limestone, and coking coal. Nucor and dozens of imitators set up so-called "mini-mills" around the country that produced a million tons of low-quality steel or less a year for use in the construction industry.

The mini-mills, most of which were built during the 1970s and 1980s, typically used nonunion labor. By the early 1980s, they had taken much of the bar, flange, and structural steel business away from integrated mills such as U.S. Steel and Bethehem. By the late 1980s, Iverson and his lieutenant, Keith Busse, built a state-of-the-art thin slab casting electric arc furnace mill near Crawfordsville, Indiana. Suddenly, Nucor

Rudy Perpich made the Iron Range sport bocce ball into a state pastime during the 1980s. Here, he instructs Duluth businessman Monnie Goldfine in some of the finer points of the game. NEMHC S3065 B2f23

Rudy Perpich worked tirelessly to ensure that state bonding dollars came to Duluth, the Iron Range and St. Louis County during the steel industry recession of the 1980s. Perpich, seen here with executives of the Duluth Chamber of Commerce, frequently was in St. Louis County to oversee pet projects. The Iron Range governor was helped in his political efforts during the decade by a state legislative delegation from northeastern Minnesota that held key committee posts in the Minnesota House and Senate. NEMHC S3065 B2f26

A youthful Herb Bergson, who has the distinction of serving both Superior, Wisconsin, and Duluth as mayor during his political career. Douglas, NEMHC S3065 B2f22

could compete with the integrated mills for the all-important automotive business.[49]

Before Nucor could begin its expansion into the thin slab market, however, the integrated side of the steel business began falling apart. Bethlehem Steel reported a $1.5 billion loss in 1982; in 1982, and 1983, steelmakers collectively lost more than $6 billion. Nearly 50 percent of the industry's capacity was shut down by 1982. Much of it would be consigned to the scrapper's torch later in the 1980s and early 1990s.[50] Employment in the Monongahela Valley, the heart of the integrated steel industry, plummeted by two-thirds between 1977 and 1982. Two years later, a third of the nation's steelworkers were unemployed.[51]

A growing number of the unemployed steelworkers were residents of St. Louis County's Mesabi Range taconite industry. For the Range, the bottom fell out in 1982. Minntac, the largest producer in the county, shipped 3.4 million tons that year, just over 20 percent of its capacity. Half of the Great Lakes fleet was laid up by mid-summer. The slightly more than 15 million tons of pellets flowing through Duluth-Superior in 1982 was the worst year for iron ore in northeastern Minnesota since the Great Depression.[52]

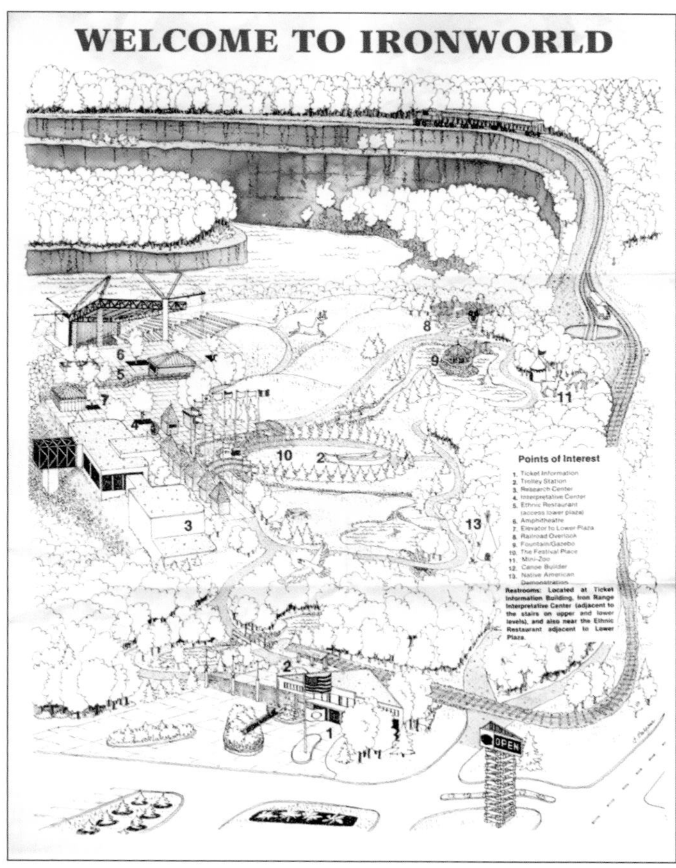

Ironworld, a tourist attraction exploring the rich heritage of iron mining in Minnesota, was opened by the Iron Range Resources and Rehabilitation Board at Chisholm in the summer of 1986. Minnesota Governor Rudy Perpich, an Iron Range native, presided over the grand opening ceremonies. NEMHC S4572, 84–89

The Lake Superior Maritime Visitors Center, opened in 1973, is one of the most popular tourist attractions in the entire Midwest. On summer afternoons, thousands of tourists visit the museum and its exhibits and line the walls of the Duluth Ship Canal to watch vessels entering and exiting the harbor. NEMHC S3065 B1f41

The human cost was staggering. More than two-thirds of the 16,500 workers employed in the Mesabi Range taconite industry were on extended layoff by the summer of 1982. Things were only marginally better in 1983 and 1984, and in some ways, even worse. Butler Taconite, one of the smaller producers, closed for good when its owner, Hanna Mining Company, was unable to secure contracts for the plant's pellets. Thousands of Iron Range residents, most of them young people low on the union seniority lists, sold out and moved away. Many of those gravitated to the booming oil industry in Texas and Louisiana.

The news only got worse in 1985 and 1986. By 1986, the U.S. steel industry had shed more than 32 million tons of capacity in a decade—the equivalent of closing one medium-sized steel mill every year since 1977. In July 1986, LTV filed for bankruptcy protection. One of the casualties was Reserve Mining Company, which promptly ceased operations.[53] Another victim of the LTV bankruptcy was Pickands-Mather, the Cleveland-based iron ore firm that had been involved in iron ore production in St. Louis County for almost a century. P-M was quickly acquired by its longtime competitor, Cleveland-Cliffs.[54]

The pain of the iron and steel contraction wasn't confined to the Range towns. Duluth was hammered by bad news during the 1970s and 1980s. U.S. Steel Corporation closed the steel plant in Morgan Park, and the federal government announced early in the decade that it was shutting down the Duluth Air Force Base in Hermantown. Jeno Paulucci, an Iron Range native who had made his first fortune with Chun King Foods, closed much of his Jeno's frozen food processing operations in the Twin Ports and transferred production to newer plants in southern Ohio.

Perhaps the most heartbreaking manifestation of the restructuring of the American steel industry—at least for residents of the county's largest city—was the dismantling of the fleet. Between 1980 and 1989, some 120 Great Lakes vessels were scrapped in their home ports or towed overseas for scrapping. Vessels that local boat watchers had grown accustomed to seeing each spring for a half-century and more were stripped and reduced to plate and structural steel for the industry's blast and electric arc furnaces.[55]

Diversification

It wasn't all doom and gloom in the county during the early 1980s. U.S. Steel, which purchased Marathon Oil in 1981, signaled that it intended to remain in the integrated iron and steel business when it spent $26 million to upgrade the company's taconite docks in West Duluth during the depths of the steel recession.[56] St. Lawrence Cement invested $18 million to build a state-of-the-art cement terminal in Duluth to distribute Canadian cement in the Upper Midwest.

Federal, state, and local governments did what they could to alleviate unemployment in the county. The completion of several massive public works projects—the Richard I. Bong Bridge between West Duluth and Superior and the Interstate 35 extension through downtown Duluth and the east end—provided employment for thousands of construction workers for more than a decade. The conversion of the Semi-Automatic Ground Environment (SAGE) building at the former Duluth Air Force Base as headquarters of the Natural Resources Research Institute (NRRI) was made possible by an innovative partnership between the state and the University of Minnesota Duluth.

St. Louis County was fortunate that it could draw upon unique resources to help surmount the economic woes of the 1980s. The IRR was a state agency chartered in 1941 to help strengthen and diversify the iron mining economy of northeastern Minnesota. The IRR teamed with the second unique resource: the man who occupied the governor's mansion for much of the 1980s.

Rudy Perpich was a prototypical Iron Ranger. Born in Carson Lake in 1928, Perpich had practiced dentistry in Hibbing before going into state politics in the 1960s. Perpich served as lieutenant governor from 1970 to 1976, and moved into the governor's office when Governor Wendell Anderson went to Washington, D.C., to take over Vice President Walter Mondale's U.S. Senate seat. Both Perpich and Anderson were defeated in 1978, and Perpich spent several years in New York City and Austria as a consultant for Minnesota-based Control Data Corporation for the next four years. In 1982, Rudy Perpich returned to Minnesota and was successful in winning the DFL primary for governor and the November election.[57]

Perpich was elected governor at a time when his beloved Iron Range was hurting, and he made economic development and diversification of the northeastern Minnesota economy a watchword of his administration. Perpich charged the IRR with using its considerable bonding powers to help diversify the area's reliance on iron mining. Perpich also convinced the IRR and the Minnesota Legislature to earmark money from taconite taxes for diversification and economic development.

IRR branched out into a number of tourism ventures, including IronWorld at Chisholm and Giant's Ridge, a ski and golf resort near Biwabik. Some of the projects, such as an ill-fated chopsticks factory in Hibbing and a proposed biomedical plant near Virginia that promised a cure for cancer, never materialized. But the Perpich and IRR initiatives helped stem the bleeding. By 1987, the region's taconite industry and the nation's steel industry had finally begun the long road back to competitive status and profitability.[58]

In 1988, Cleveland Cliffs produced 40 million tons of pellets, most of them from the Mesabi Range and the best performance since 1982.[59] Other steel firms would fail in the years ahead, and the steel industry contraction would require fewer pellets to feed the blast furnaces. Half the jobs the taconite industry had reported in 1979 existed a decade later. But St. Louis County and the Range would come back during the 1990s as the nation's primary source of iron ore.

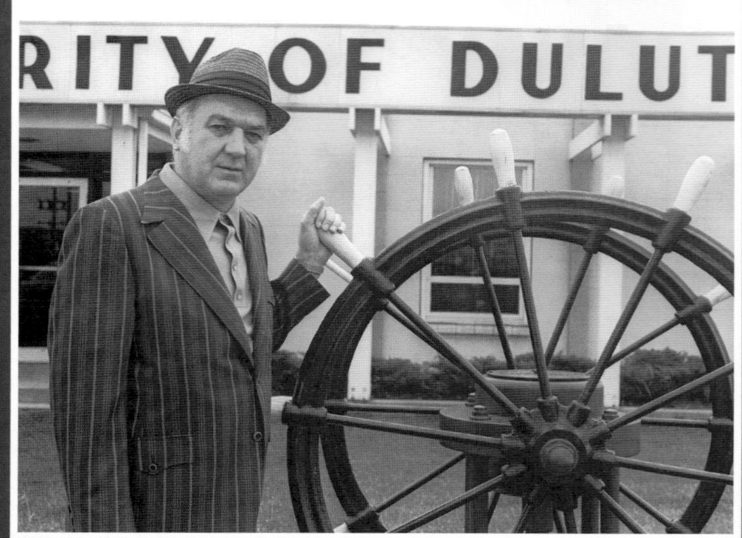

Tony Rico's trademark porkpie hat was a familiar sight around the waterfront during the 1970s and 1980s. Rico, a native of Duluth's Raleigh Street Italian neighborhood, had so many business interests in the port—including control of the maritime pilots who guided saltwater vessels into the Twin Ports—that wags often referred to the Port of Duluth as "Port-o-Rico." NEMHC S3065 B2f26

Computerization in St. Louis County was in its infancy in the late 1970s and early 1980s, but use of computers in both the public and private sectors would explode during the 1980s. NEMHC S3065 B2f23

John Fedo, second from right, confers with officials from the Duluth Chamber of Commerce and the Duluth Convention and Visitor's Bureau early in his first term as mayor of Duluth. Fedo spent most of the 1980s as mayor of the county's biggest city. Although he exited office in the midst of a political scandal, Fedo directed the conversion of the city's Canal Park waterfront from an industrial wasteland to one of the Midwest's top tourist destinations. Fifield Powers, NEMHC S3065 B2f23

The size of the open pits dotting the Mesabi Iron Range is such to boggle the mind. During the 1980s, the IRR worked hard to create reclamation models so that residents in future generations might enjoy the recreational benefits of living beside a chain of huge lakes along what once was the Mesabi Range. NEMHC S3065 B2f5

Writers, editors, and graphics artists found employment in Duluth during the 1980s with Ojibway Press, a longtime local publisher of trade magazines. In the 1990s, the firm was acquired by East Coast investors, renamed Advanstar, and much of the editorial function of the magazines was moved from Duluth. NEMHC S3065 B2f22

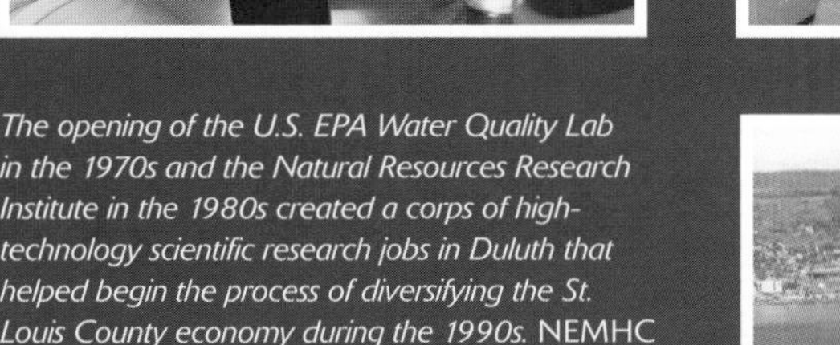

The opening of the U.S. EPA Water Quality Lab in the 1970s and the Natural Resources Research Institute in the 1980s created a corps of high-technology scientific research jobs in Duluth that helped begin the process of diversifying the St. Louis County economy during the 1990s. NEMHC S3065 B2f23

The 1976 opening of the Ortran coal facility on the Superior side of the harbor created an industry that would diversify the Twin Ports' traditional reliance on iron ore bulk cargoes. Landmark clean air legislation passed in the early 1970s mandated electric utilities to burn low-sulfur coal in steam electric generating plants. Minnesota Power had pioneered the burning of low-sulfur coal hauled by rail from Montana to northern Minnesota in the late 1960s. In 1976, Detroit Edison took the idea one step further, building a coal terminal facility at Superior so it could load vessels with the coal for shipment to power plants on Lake St. Clair north of Detroit. Critics laughed at the idea, but a quarter-century later, the renamed Superior Midwest Energy Terminal was moving 20 million tons of coal a year through the terminal. NEMHC S3065 B1f50

The snowmobile transformed the tourism business in rural St. Louis County, northeastern Minnesota, and the Upper Great Lakes states. From World War II until the 1980s, resorters in the region were open from May to October, primarily for fishing season and the fall color time. Resorters had a difficult time in justifying capital investment in an asset that could only produce income a little over four months a year. With the introduction of the snowmobile, however, February could be as profitable a month as August. St. Louis County groomed trails through county- and state-owned land. Snowmobilers from the Twin Cities, southern Minnesota, and Iowa had a new winter playground to run their machines—and spend money all winter long. NEMHC S3065 B2f3

Duluth's hills overlooking Lake Superior during the 1980s began to attract legions of young people from the Twin Cities interested in testing their mettle in bicycle races. Mountain biking in Duluth and in hilly northern communities such as Tower-Soudan and Ely became hugely popular during the 1990s. NEMHC S3065 B2f1

Will Steger and Paul Schurke, center, focused the world spotlight on St. Louis County in 1986 when they staged their assault on the North Pole from a base camp in Ely. Steger and his team of intrepid adventurers used dog sleds in their unsupported run for the Pole. They were financially supported by numerous St. Louis County businesses and residents. One of the members of the team, Ann Bancroft, was the first woman to gain the Pole using dogs.
NEMHC S3065 B2f27

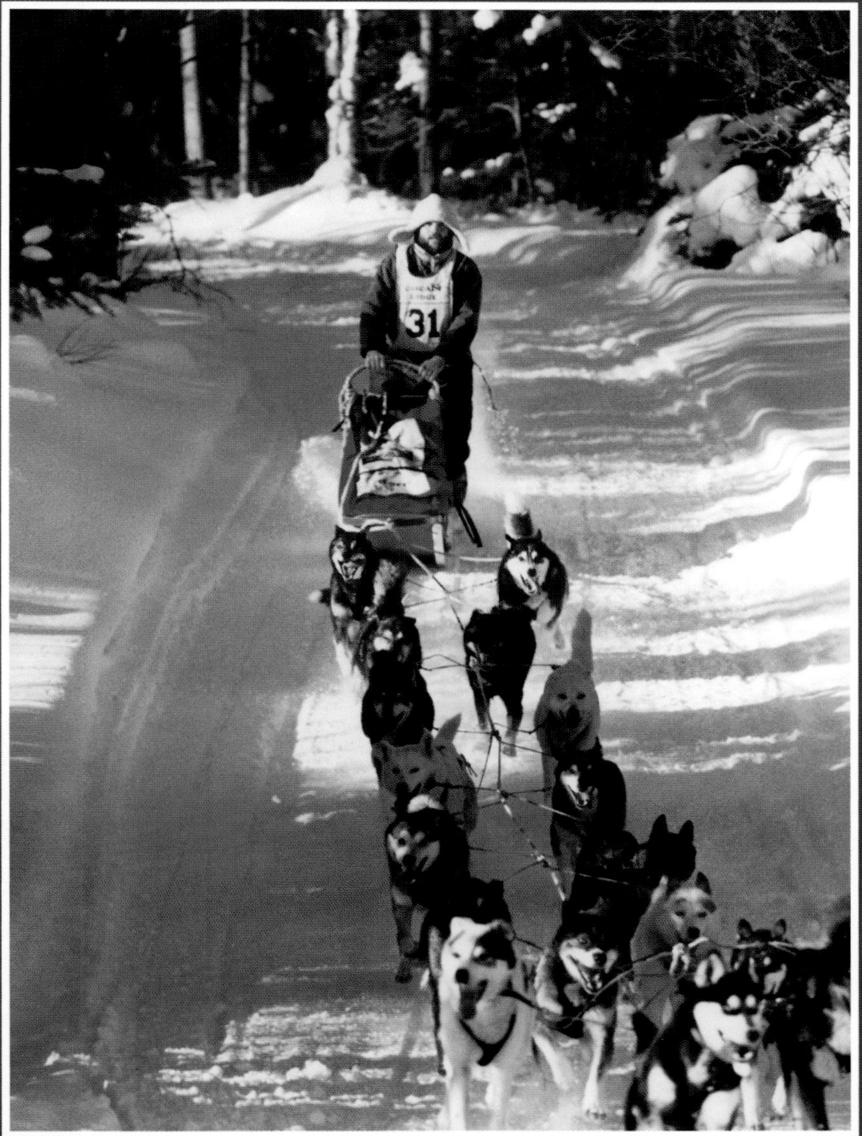

The Steger International Polar Expedition trained in Ely in 1986. Will Steger and Paul Schurke led a team that used dog sleds to reach the North Pole, much the same way Commander Robert Peary gained the Pole seventy-some years before. NEMHC S3065 B2f27

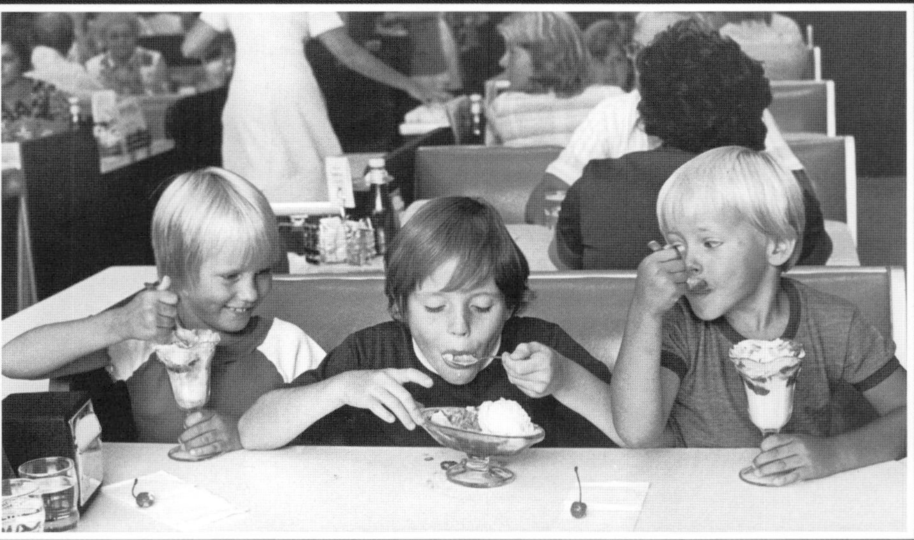

Youngsters enjoy ice cream treats at Bridgeman's, a chain of northern Minnesota dairy stores that was well represented in St. Louis County since 1905. Like youngsters everywhere, these three St. Louis County residents are saving the maraschino cherry for last. NEMHC S3065 B2f22

Rudy Boschwitz teamed with Dave Durenberger to keep Minnesota's two U.S. Senate seats safe for the Republican Party during the 1990s. A successful homebuilding store executive, Boschwitz worked well with Congressman James Oberstar and Governor Rudy Perpich to represent the interests of St. Louis County and northeastern Minnesota. NEMHC S3065 B2f22

Dennis Anderson is one of the more familiar faces of St. Louis County since the 1980s. As anchor for WDIO-TV in Duluth and its ABC-TV Iron Range affiliate, Anderson is welcomed into thousands of homes every evening to report the day's events. Anderson still signs off the air each evening with the tagline, "Be kind." NEMHC S3065 B2f22

Arne Carlson, shown here with former Duluth Mayor Ben Boo, defeated Rudy Perpich in the 1990 gubernatorial election. Carlson had been defeated in the Independent-Republican primary, but when the party's candidate was caught skinny-dipping with a teen-ager, Carlson was drafted to run against Perpich as a last-minute replacement. NEMHC S3065 B2f15

Snowmobiles and ice fishing became winter recreation staples in the county during the 1970s. The popularity of snowmobiling, ice fishing, and cross-country skiing became so pronounced during the 1990s that some resorts in the Ely area were remaining open year-round. NEMHC S3065 B1f27

Sailors tack into the wind on Lake Superior on a bright summer afternoon. Sailing has become increasingly popular in the region, following the construction in the 1980s of major new small craft harbors at Superior and Bayfield, Wisconsin. NEMHC S3065 B1f6

chapter seven

Recording the Heritage

The St. Louis County Historical Society, 1922–2006

The people of St. Louis County have always had an abiding interest in the history of their region. For more than a century, county residents have banded together to study the rich past of Minnesota's largest county.

Today, the residents of St. Louis County accomplish the worthy goal of understanding the local past through the St. Louis County Historical Society. A county-funded, member-supported nonprofit corporation, the Society collects, presents, and preserves knowledge of the history and prehistory of both the county and region. The Society operates a museum and research archives facility as well as providing operational support to six operating affiliates.

Although not formed officially until 1922, the Society's membership has its origins in earlier groups. The oldest was formed in 1884 to recognize those who had first arrived at the Head of the Lakes. The Old Settlers Association of the Head of Lake Superior was founded in Superior and required proof that prospective members had lived in the Twin Ports for at least twenty-five years. In the first year of membership, only those who had arrived in Duluth or Superior before the Civil War qualified.

The Old Settlers Association was somewhat exclusive through the turn of the twentieth century. Duluth and Superior had registered a population boom in 1869 and the early 1870s when the '69'ers arrived, but most had left in the wake of the Panic of 1873. Consequently, Twin Ports residents with twenty-five years of residency didn't begin to register appreciable numbers until after 1900. But the Old Settlers Association did kindle an interest in the region's history and also spawned similar organizations elsewhere in the county. The First Settlers Association of Hibbing

The people of St. Louis County have always had an abiding interest in...

St. Louis County Historical Society rooms/office in St. Louis County Courthouse. At left, William E. Culkin, first St. Louis County Historical Society President. Alice Lindquist, at right, was part of the support staff in 1926. NEMHC S2386 B33f11

was established in 1893, an Old Settlers Association for the Mesabi Range shortly after, and the Vermilion Range Old Settlers in 1915.

In 1897, a small group of Duluth residents proposed an organization for local people interested in the natural history of St. Louis County and northeastern Minnesota. They named their organization the Historical and Scientific Association of Duluth, Inc., and announced the formation of membership sections that dealt with history, zoology, botany, anthropology, geology, and mineralogy, but were soon incorporated into the earlier Old Settlers Association of Lake Superior.

The Society's primary predecessor was the Minnesota War Records Commission. Authorized in 1918 by the Minnesota Commission on Public Safety to "insure a proper historical compilation of activities and units" of the Great War, the Commission had chapters in each county. St. Louis County's branch was headed by William E. Culkin, a prominent Duluth attorney who had arrived in Duluth in 1896 to serve as a registrar in the United States Land Office.[1] Culkin's leadership of the War Records Commission carried over into a broader mandate to preserve the county's history.

The Duluth Convention

A major interest in the county's history came in support of the 1921 publication of the American Historical Society's three-volume *Duluth and St. Louis County, Minnesota: Their Story and People.* Edited by Walter Van Brunt, the book was funded by the hundreds of country residents who told their life stories to the publishing company's interviewers. Van Brunt also drew upon the collected memories of dozens of members of the Old Settlers Associations in both the Twin Ports and on the Iron Range.

In July 1922, Solon J. Buck, the influential superintendent of the Minnesota Historical Society, traveled to Duluth for a Society "historical convention." The convention was to introduce the people outstate to the work of the Society and proposed outreach activities.

At an informal luncheon, Buck outlined to Culkin and others present his hope for the creation of a branch society in St. Louis County that would be affiliated with the Minnesota Historical Society. Culkin also was aware that legislation had been introduced in the Minnesota General Assembly to allow the County Board of Commissioners to support a county historical society with an appropriation of up to $2,500 a year.[2]

Culkin invited forty-six prominent county residents—most were Duluthians—to attend an organizational meeting on November 18, 1922. The list contained some of the most influential names in the county: Ensign, Congdon, Alworth, Whiteside, Merritt.[3] The meeting decision was unanimous in its desire to form a county historical society. One week later, at the Duluth Chamber of Commerce offices, the group met again to formally organize the St. Louis County Historical Society and adopt a constitution. Culkin was elected the Society's first president. On December 1, 1922, he presided over the first regular meeting of the Society.

Early Years

William E. Culkin presided over the affairs of the St. Louis County Historical Society for fifteen years as both elected president and paid staff director. The St. Louis County Commission made an office available for the Society at Room 410 in the County Courthouse and appropriated $2,500 for Society activities beginning in 1924. Later, Culkin was able to store World War I service records and maps in Room 507 and in an attic room on the sixth floor.

Membership during the Culkin years peaked at 187 in 1931, a level it wouldn't reach again until after World War II. But Culkin, and an early successor as president, Richard Bardon, guided the fortunes of the Society for almost forty years. During that period of time, Culkin, Otto Wieland, who served as president from 1937 until his untimely death in 1943, and

St. Louis County Histrical Society Museum, 2228 East Superior Street, Duluth, 1947–1976. NEMHC S2386 B33f11

Bardon, who was president from 1943 to 1961, scheduled annual meetings open to the public that interspersed musical entertainment with the reading of papers on the county's history. As early as 1924, Culkin established a policy of moving the Society's meetings to other communities in St. Louis County when he hosted an annual meeting in Hibbing.

In 1943, during World War II, the Society moved its offices from the County Courthouse to two rooms in Tweed Hall on the campus of the Duluth State Teachers College. Space limitations at the Courthouse had stalled the Society's plans for a historical museum, and the Society would occupy space in Tweed Hall—the former J. B. Cotton residence on East First Street—until 1946.

The postwar years resulted in a membership boom, doubling it from 258 in 1947 to 521 in 1953. Space in Tweed Hall was cramped by 1946, and when the St. Louis County Board of Commissioners offered the Society a tax-forfeit mansion at 2228 East Superior Street, Bardon eagerly accepted. The Society moved into the former George and Marian Stone mansion in 1947 and immediately began renovating the seventeen-room building with its seven fireplaces into offices and a museum. The Society would occupy the Stone mansion for the next thirty years.

The postwar years also produced a major organizational change. In 1951, the Society was formed on a nonprofit basis, and seven years later, in 1958, the membership voted to incorporate the Society as a nonprofit corporation.[4] Early in 1956, the year that Albert Woolson, the last Union Army veteran of the Civil War, died in Duluth, the board was increased to twelve members and a membership in the Society was $1.[5]

During the 1950s and 1960s, the Society continued public programs, including a presentation by renowned Minnesota historian and author Grace Lee Nute at the 1952 annual meeting in Duluth. In 1962, the Society successfully petitioned the U.S. Board of Geographic Names to use the name Hearding Island for an island in Superior Bay to honor the memory of William Hearding, who first surveyed the Duluth-Superior harbor in 1861.

In 1954, Elizabeth deCourcy Greene was the Society's first staffer called by the title executive secretary. Greene ran the day-to-day affairs of the Society from 1954 to 1967.

Beginning with Greene, the Society had an executive employee separate from the organization's president. Arthur O. Roberts, who succeeded Richard Bardon as president in 1961 and capably guided the fortunes of the Society through the remainder of the decade, estimated in 1966 that some six thousand visitors toured the Society's museum.[6] But as the 1960s gave way to the 1970s, the Society was once again wrestling with the problem of space constraints.

The Dawn of the Modern Era

The Society's membership of about five hundred paying members was static during the decade. Greene was succeeded in 1967 by Arnold Luukkenon, who was succeeded in 1968 by Frank A. Young, who would be executive secretary until 1977.

Young's tenure was marked by two events that would characterize the Society's operations for the next thirty years. By 1976, the Society had collected historical papers for more than a half-century. Few were organized or cataloged; most were stored in window boxes, attics, and closets of the Society's house mansion. The Minnesota Historical Society (MHS) strongly urged the Society to partner with them and the University of Minnesota Duluth (UMD) to provide for the collections' care and make them accessible for researchers.

In 1976, MHS and UMD established a regional research facility called the Northeast Minnesota Historical Center at the UMD Library. The Society joined the partnership the next year when it deposited its research collection at the Center.

The second major event for the Society came in 1976 when it sold its longtime home and moved into a new facility in downtown Duluth. In 1976, the St. Louis County Board appropriated $63,650 for Society operations, but indicated funds would be substantially less for 1977. County funding for the Society was consistent but cyclical and dependent upon the financial health of the state. To offset the expected funding drop, the Society's board put the Superior Street house up for

St. Louis County Historical Society exhibit with MHS 1998–1999, Jewish Life in the Northland. NEMHC S2386 B33f11

sale and voted to move into the St. Louis County Heritage and Arts Center, the renovated 1892 Union Depot, at 506 West Michigan Street. The Society would occupy quarters at the Heritage and Arts Center for the next thirty years.

Frank Young was succeeded in 1978 by Larry Sommer, a Carleton College graduate who would coordinate the Society's affairs until 1989. Sommer and the Society's board joined the Lake Superior Museum of Transportation and the Duluth Children's Museum in 1982

to open Depot Square, an exhibit of 1910 replica, half-size Duluth business storefronts on the train museum level of the Heritage and Arts Center. During 1982, total paid attendance at the Center was almost 137,000.

In 1983, the Society was responsible for publishing two books. Glenn Sandvik, an editorial writer for the *Duluth News-Tribune,* wrote *Duluth: An Illustrated History of the Zenith City,* published by Windsor Publications. Patricia Johnston's *Eastman Johnson's Lake Superior Indians,* published by Afton Press, illustrates the thirty-two Johnson oil paintings and drawings in the Society's collection.[7]

The early 1980s was also an era of change for the Society. In 1982, a change in state law provided that more than one historical society in a county may receive funds. Since then, the Society re-grants funds to six affiliated historical societies on the Iron Ranges, including the Hibbing Historical Society, Sisu Heritage, Minnesota Museum of Mining, Ely-Winton Historical Society, Virginia Area Historical Society, and Tower-Soudan Historical Society.

Membership grew to more than seven hundred people by the mid-1980s, just two years after the Society's board authorized a restricted endowment fund in 1983.[8] The Depot Free Day in February 1983 was the best ever attended, with fifty-eight hundred visitors passing through the doors. The Society's volunteers came into their own during the 1980s, donating more than sixty-five hundred hours in 1985 alone.[9] In 1987, volunteers assisted Society staff in recataloging artifacts and storing them in acid-free boxes.[10]

The Coombe Years

In the summer of 1989, Larry Sommer accepted the post as the head of the Montana State Historical Society in Helena.[11] The board quickly appointed JoAnne Coombe and Maryanne C. Norton co-administrators of the Society. When Norton resigned in the fall of 1992, Coombe became the Society's fulltime executive director.

An Ely native and graduate of Vermilion Community College, Coombe presided over several major developments during the 1990s and into the twenty-first century. In 1990, the Forest History Gallery—formerly the J. C. Ryan Forest History Room—was developed. The following year, the Society developed the Fesler Gallery with its Herman Melheim Collection of carved furniture.[12] In 1994, the Society formed an American Indian Advisory Committee and announced a $100,000 grant to renovate the Sieur duLhut Room and rename the room the Lake Superior Ojibwe Gallery in 2000.[13]

For much of the late 1990s and early twenty-first century, the Society helped develop the World War II Living History Project and the Veterans Memorial Hall at the St. Louis County Heritage and Arts Center. Partially funded with a $175,000 State of Minnesota Challenge Grant, the Veterans Memorial Hall web site went on line in 2001 to chronicle the lives of northern Minnesota veterans of conflicts from the Civil War to the Gulf War.[14] *Dignity through Unity,* a Vietnam exhibit, opened in April 2004 for a three-year run. The exhibit was recognized for excellence in 2005.

The Society celebrated its eightieth anniversary at the Kitchi Gammi Club in Duluth on October 8, 2002, and celebrated the twentieth anniversary of Depot Square that same year. Today, the Society continues to collect historically and culturally significant artifacts and documents pertaining to the county's past. The Society operates both a state-of-the-art museum and reaches out to citizens countywide through a network of Iron Range affiliates. St. Louis County appropriates more than $210,000 for Society general operations, serving more than 120,000 individuals each year through museum, archives, web site, and affiliate activities.

William Culkin would undoubtedly be proud of what he set in motion back in 1922.

Epilogue

As St. Louis County celebrates its sesquicentennial anniversary in 2006, the natural resources economy that has long been the foundation of the county's prosperity or despair is once again in an upswing. After barely surviving the complete reengineering of the North American steel industry in the 1980s, the Mesabi Range and the Port of Duluth-Superior recovered in strong fashion during the 1990s. Shipments of taconite pellets to the remaining, far more productive blast furnaces and mills on the Lower Great Lakes rebounded during mid-decade to some of the best yearly levels since before the industry's sudden collapse in the early 1980s.

Employment levels in the mines and mills of the Mesabi and the rail and vessel infrastructure of the Twin Ports never recovered to the levels of 1979. But because of automation and productivity gains, those who did find work in the region's iron industry were well compensated.

Changing environmental regulations and traffic patterns brought new cargoes to the Twin Ports, a community that has always looked to its waterfront for sustenance. When Congress mandated tough new clean air regulations in the early 1990s, steam electric utility generators in the Midwest were forced to convert to low-sulfur coal from Montana, Wyoming, and Colorado for their boilers. The most cost-efficient way to get coal from the Great Plains to utilities in Michigan and Ohio was by rail to the Head of the Lakes and then by one-thousand-foot ore carriers to the steam electric-generating plants on the Lower Lakes. By the late 1990s, more than 20 million tons of low-sulfur Western coal were moving through Duluth-Superior each season, an amount undreamed of ten or fifteen years before.

The renewed prosperity enjoyed by the iron and shipping industries made Duluth a regional center for education, health care, and retail sales. The Zenith City's old-line hospitals, St. Luke's and St. Mary's, entered into alliances with physicians' groups and the Duluth Clinic to provide comprehensive medical care not only to Duluth and Superior but to the entire region. The St. Mary's–Duluth Clinic (SMDC) alliance began moving into the Iron Range and adjacent regions of northern Minnesota, competing with home-grown health care establishments that had been caring for Rangers since the days of the Merritts, a century earlier.

University of Minnesota Duluth (UMD) became an engine driving the Duluth and regional economy. By the late 1990s, the campus overlooking Lake Superior boasted its own professional schools and a growing enrollment of students from the suburbs of the Twin Cities. The University's Natural Resources Research Institute (NRRI) gave UMD a developing reputation for cutting-edge research in how best to utilize the natural resources that Mother Nature has bestowed on the region.

Tourism continued to emerge as a pillar of the region's economy. In less than fifteen years, Duluth converted an industrial wasteland between downtown and the Duluth Aerial Bridge into Canal Park, one of the most popular tourist destinations in the Upper Midwest. On soft summer evenings in the 1990s and 2000s, thousands of Twin Citians, Iowans, and Hoosiers thronged Canal Park and its myriad hotels, fashionable restaurants, and trendy shops, drawn to the area by the mournful blasts of a saltwater vessel's horn, answered by the bridge tender's whistle.

The federal government and the state did their part to expedite the flow of traffic to St. Louis County and the North Shore of Lake Superior. An ambitious Interstate

Highway extension project took more than a decade to complete, but when it was finished, I-35 emerged from tunnels beneath the city of Duluth into the many moods of the rockbound Superior shore.

The original west wing of St. Luke's Hospital, about 1925. McKenzie, NEMHC S2422 B1, no. 8045

Iron Range Resources (IRR) made tourism a key component of its efforts to diversify the Iron Range from an over-reliance on taconite. IRR's Giant's Ridge project in the northeastern part of the county grew to encompass ski hills, golf courses and condominium developments, all in less than a decade.

Another tourism initiative that became increasingly common during the 1990s was the expansion of Native American gaming. The City of Duluth pioneered the concept in the 1980s when it joined the Fond du Lac band of Anishinabe from nearby Cloquet in refurbishing an abandoned Sear's store on East Superior Street to house bingo and slot machines. The Fond du Lac band later developed the major Big Bear casino complex on I-35 just south of Duluth, and the Bois Fort band opened a smaller casino on tribal land near Lake Vermilion.

Economic development efforts, some dating back to the collapse of the steel industry in 1980, began to pay dividends. Cirrus, a start-up private plane manufacturer, took up residence at the Duluth International Airport and began producing what many aviation experts considered the finest, most easily maneuverable private aircraft made in the world. By the early 2000s, the company had a long waiting list for its product. Two or three times a week, Mesaba Aviation flights from the Twin Cities transported a passenger who was coming to the Twin Ports from California or Maryland or Texas to pick up his or her Cirrus aircraft.

Retail in the county experienced a sharp recovery during the 1990s. Big-box retailers increasingly made the Highway 53 corridor in the Miller Hill area of Duluth the shopping destination for much of St. Louis and surrounding counties. During the 1990s, increasing numbers of Canadians from Thunder Bay journeyed to Duluth's Miller Hill retail concentration to avoid paying the high value-added tax on clothing and consumer goods in their native Ontario. Hibbing and Virginia also developed a sharply expanded retail presence, but it was often located away from the Range communities' traditional downtowns.

Global Reach

The prosperity of the 1990s, reminiscent of the growth of the county at the turn of the previous century, was accompanied by several worrisome trends that presaged difficulties for St. Louis County. Global competition was perhaps the pre-eminent trend in the economy during the 1990s and a real threat to the continued viability of the region's iron mining economy. And the steady loss of prestige and influence of organized labor in American industry complicated the global threat to the domestic iron and steel sector.

The Iron Range had long understood the ramifications of the global economy. As early as the mid-1980s, there were concerns in the Lake Superior taconite industry that natural iron ore from Brazil could be landed at Lower Lake Michigan ports for less than Mesabi Range pellets. The concern was unfounded, but by the late 1990s, millions of tons of semifinished steel slabs from overseas were being landed at American ports. Each ton of foreign slab eliminated the need for about 1.5 tons of domestic iron pellets, which kept Mesabi Range taconite plants from expanding, at the very least.

The foreign slabs were accompanied by as many as 25 million tons of imported finished steel each year in the late 1990s and early 2000s. The result was a wave of steel company bankruptcies.

The industry's problems at the dawn of the new millennium weren't solely because of global competition. The industry shakeout at the time was as much due to the growth of minimills as it was to a perceived flood of imported steel. Minimills operated by companies such as Nucor and SDI continued to expand. By 2000, the minimill sector, which made steel in electric arc furnaces from scrap, accounted for about half of American steel production.

Integrated producers, such as Wheeling-Pitt, LTV, and Bethlehem, which made steel in blast furnaces from iron pellets, were the real victims of foreign competition. Saddled with often outmoded facilities and struggling to pay growing pension costs, they began to seek federal bankruptcy protection.

By the mid-1920s, St. Mary's Hospital had a capacity of 285 patients, most housed in the new wing (back left), which opened in the summer of 1922. Florman, NEMHC S2386 B7f20

The original wing of St. Mary's Hospital, shortly before the turn of the twentieth century. NEMHC S2386 B7f20

The troubles in the steel industry rippled to St. Louis County and the Mesabi Iron Range. Between 2001 and 2003, both Erie Mining Company and Eveleth Taconite Company closed when their owners either filed for bankruptcy protection or got out of the steel business. National Steel Pellet Company was sold to U.S. Steel Corporation. The Pittsburgh-based steel giant even talked about exiting the taconite business itself, and U.S. Steel did sell its USS Great Lakes Fleet and Duluth, Missabe and Iron Range subsidiaries.

The closure of Erie and Evtac sent shock waves through the Iron Range, and many predicted that the 110-year-old industry would soon be but a memory in St. Louis County.

Evidence that global competition was having a fundamentally transforming impact on the American economy wasn't hard to uncover in northern Minnesota during the late 1990s and early 2000s. The region's pulp and paper industry, long a staple of employment in small-town northern Minnesota, went through wrenching change during the period.

When the downsizing and ownership changes were complete, the industry essentially was foreign-owned. Mills in places such as Grand Rapids and Cloquet, which had long employed St. Louis County residents, were suddenly owned by multinational forest products giants in Finland and South Africa. The Lake Superior Paper Industries in West Duluth was acquired by Stora Enso of Sweden.

But what global economic forces took away from northern Minnesota, they also could confer benefits on the region. After the turn of the millennium, China emerged as a developing superpower. Chinese demand for steel in the early twenty-first century was insatiable. The Chinese steel industry could not find enough iron from Australia and Brazil, its traditional suppliers, to feed the country's blast furnaces.

Working with the office of Congressman James Oberstar, Chinese steelmakers and Cleveland-Cliffs, owners of some 40 percent of the nation's iron reserves, forged an innovative agreement in 2003. The pact called for China's Laiwu Steel to join Cleveland-Cliffs in a joint venture to purchase, reopen, and operate Evtac. At the same time, Minntac was shipping about 1 million tons of taconite pellets from Mountain Iron to China. For the first time in history, the Mesabi Range's production was global in scope.

The increased global focus did not stop with Chinese ownership of Range producers. As the steel industry rebounded sharply in 2004 and 2005, an increasing share of U.S. ownership consolidated overseas. Entrepreneur Wilbur Ross, whose ISG Corporation had turned around Bethlehem Steel, sold his iron and steel holdings to Mittal Steel Corporation, a European firm with roots in India. With the closure of the Mittal acquisition, Hibtac and Minorca joined Evtac as foreign-owned entities.

Through it all, St. Louis County had endured. Since it was first hacked out of the wilderness, the county and its residents have exhibited a resilience and persistence that have allowed the region to survive and thrive in the ups and downs of a natural resource-based economy.

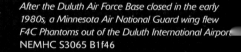

After the Duluth Air Force Base closed in the early 1980s, a Minnesota Air National Guard wing flew F4C Phantoms out of the Duluth International Airport.
NEMHC S3065 B1f46

Notes

Chapter 1

1. Douglas A. Birk, "When Rivers Were Roads: Deciphering the Role of Canoe Portages in the Western Lake Superior Fur Trade," in Jennifer S. H. Brown, W. J. Eccles, and Donald P. Heldman, eds., *The Fur Trade Revisited* (East Lansing: Michigan State University Press, 1994), p. 359.

2. Ibid., p. 361.

3. *Building a Birchbark Canoe*, www.home.achilles.net/~judyk/birchbarkcanoe/toolspage/tools.htm.

4. Walter Edmonds, *The Musket and the Cross* (Boston: Little Brown and Company, 1968), pp. 35–41.

5. Daniel Greysolon, Sieur Du Lhut, *The Catholic Encylopedia*, www.newadvent.org/cathen/05188a.htm.

6. Ibid.

7. Bill Beck and C. Patrick Labadie, "Ahmik" in *Welcome to Pride of the Inland Seas: Companion Stories*, www.duluthport.com/pride/ahmik/html.

8. Peter C. Newman, *Empire of the Bay: The Company of Adventurers that Seized a Continent* (New York: Penguin Books, 1998), pp. 43–44.

9. Bernard Grun, ed., *The Timetables of History: A Horizontal Linkage of People and Places* (New York: Touchstone Books, 1982) pp. 348–353.

10. Newton H. Winchell, "Minnesota's Northern Boundary," *Minnesota Historical Society Collections* 8 (1898): 185–212.

11. Thomas LeDuc, "The Webster-Ashburton Treaty and the Minnesota Iron Ranges," *Journal of American History* 51 (December 1964): 476–481.

12. Grace Lee Nute, "Hudson's Bay Company Posts in the Minnesota Country," *Minnesota History* 22 (September 1941): 270–289.

13. Nute, "Posts in the Minnesota Fur Trading Area," *Minnesota History* 11 (December 1930): 353–385.

14. Frank E. Ross, "The Fur Trade of the Western Great Lakes Region," *Minnesota History* 19 (September 1938): 271–307.

15. *Time Line—A Brief History of the Fur Trade*, http://www.whiteoak.org/learning/timeline.html.

16. Newman, *Empire of the Bay*, p. 45.

17. *Time Line—A Brief History of the Fur Trade.*

Interlude: 1856

1. William Watts Folwell, *A History of Minnesota*, v. 1 (St. Paul: The Minnesota Historical Society, 1956), p. 352n.

2. Harlan Hatcher, *A Century of Iron and Men* (Indianapolis: The Bobbs-Merrill Company, Inc., 1950), pp. 23–25.

3. William B. Gates Jr., *Michigan Copper and Boston Dollars* (Cambridge: Harvard University Press, 1951), p. 3.

4. Angus Murdoch, *Boom Copper* (Ann Arbor: Copper Country Edition, 1970), p. 26.

5. Hatcher, *A Century of Iron and Men*, pp. 47–55.

6. David J. Krause, *The Making of a Mining District: Keweenaw Native Copper, 1500–1750* (Detroit: Wayne State University Press, 1992), p. 212.

7. Edmund Jefferson Danziger Jr., *The Chippewas of Lake Superior* (Norman: University of Oklahoma Press, 1978), pp. xvi–xvii.

8. Ibid., pp. 4–5.

9. Walter Van Brunt, *Duluth and St. Louis County, Minnesota: Their Story and People*, v. 1 (Chicago and New York: The American Historical Society, 1921), pp. 59–60.

10. Shiela Reaves, *Wisconsin: Pathways to Prosperity* (Northridge, California: Windsor Publications, 1988), p. 58.

11. Rachael Martin, "North Country History: Duluth's Early Churches," *The Senior Reporter* (April 2005): 10.

12. Quoted in Larry Lankton, *Beyond the Boundaries: Life and Landscape at the Lake Superior Copper Mines, 1840–1875* (Oxford: Oxford University Press, 1997), p. 33.

13. Walter Havighurst, *The Long Ships Passing: The Story of the Great Lakes* (New York: Macmillan Company, 1942), p. 203.

14. Catherine Calhoun, "Through the Locks," *American Heritage* (September 1992): 26.

15. Glenn N. Sandvik, *Duluth: An Illustrated History of the Zenith City* (Northridge, California: Windsor Publications, 1983), p. 17.

16. Rachael Martin, "North Country History: Duluth's First Church Originated in Superior," *The Senior Reporter* (February 2005): 8.

17. Van Brunt, *Duluth and St. Louis County*, v. 1, p. 126; *See also* Minnesota Historical Society, Peet Diary, 1856.

Chapter 2

1. Steve Fraser, *Every Man a Speculator: A History of Wall Street in American Life* (New York: HarperCollins, 2005), p. 74.

2. Quoted in Van Brunt, *Duluth and St. Louis County*, v. 1, p. 137.

3. Ibid., p. 137.

4. *The City Grows*, www.inredllc.com/duluth/stories/grows.htm.

5. Quoted in Van Brunt, *Duluth and St. Louis County*, v. 1, p. 139.

6. Ibid., pp. 145–148.

7. Van Brunt, pp. 163, 343; *See also* Sandvik, *Duluth: An Illustrated History of the Zenith City*, p. 25.

8. David A. Walker, *Iron Frontier: The Discovery and Early Development of Minnesota's Three Ranges* (St. Paul: Minnesota Historical Society Press, 1979), p. 21.

9. Ibid., p. 22.

10. *Wall Street History, Jay Cooke, Part II*, www.buyandhold.com.

11. Walker, *Iron Frontier*, p. 24.

12. Fraser, *Every Man a Speculator*, p. 122.

13. Ibid., pp. 122–123.

14. Davis Helberg, Maritime Day Speech, Duluth, Minnesota, 2001, n.p.

15. Steven J. Wright, "The Forgotten November Storm," *The Nor'Easter* 12, no. 5 (September–October 1987): 1–3.

16. J. D. Ensign, *History of Duluth Harbor* (Duluth: 1898), p. 3.

17. Duluth-Superior Harbor Statistics, Duluth Superior Port Authority, 1999, p. 1.

18. U.S. Army Corps of Engineers, Chronology of Harbor Improvements, 1940, pp. 3, 6.

19. Duluth-Superior Harbor Statistics, p. 2.

Chapter 3

1. Hiram M. Drache, *Koochiching: Pioneering Along the Rainy River Frontier* (Danville, Ill.: The Interstate Printers and Publishers, Inc., 1983), p. 16.

2. David A. Walker, *Iron Frontier: The Discovery and Early Development of Minnesota's Three Ranges* (St. Paul: Minnesota Historical Society Press, 1979), p. 24.

3. Ibid.

4. Ibid., p. 27.

5. Stewart Holbrook, *Iron Brew: A Century of American Ore and Steel* (New York: Macmillan, 1937), pp. 206–209.

6. Hatcher, *A Century of Iron and Men*, pp. 148–150.

7. Marvin Lamppa, *Minnesota's Iron Country: Rich Ore, Rich Lives* (Duluth: Lake Superior Port Cities, Inc., 2004), pp. 46–47.

8. Ibid., p. 48.

9. Walker, *Iron Frontier*, p. 30.

10. Lamppa, *Minnesota Iron Country*, p. 57.

11. Ibid., p. 58.

12. Ibid., p. 59.

13. Upham, p. 490; *See also* Walker, *Iron Frontier*, p. 62.

14. Lamppa, *Minnesota Iron Country*, p. 65.

15. Ibid., p. 66.

16. Beck and Labadie, *Pride of the Inland Seas*, p. 65.

17. Ibid., p. 65.

18. David A. Walker and Stephen P. Hall, Duluth-Superior Harbor Cultural Resources Study (St. Paul: St. Paul District, U.S. Army Corps of Engineers, 1976), p. 76.

19. Holbrook, *Iron Brew*, pp. 100–101.

20. Ron Chernow, *Titan: The Life of John D. Rockefeller, Sr.* (New York: Random House, 1998), pp. 382–383.

21. Ibid., pp. 383–385.

22. Walker and Hall, Duluth-Superior Harbor Cultural Resources Study, p. 78.

23. Ibid., p. 78.

24. Duluth-Superior Harbor Statistics, p. 2.

Interlude: 1906

1. Lamppa, *Minnesota's Iron Country*, p. 162.

2. Fred J. Thompson, Production Director, *Iron Range Country: A Historical Travelogue of Minnesota's Iron Ranges* (Eveleth, Minn.: Iron Range Resources and Rehabilitation Board, 1979), p. 104.

3. Thompson, *Iron Range Country*, p. 165.

4. E-mail, Pat Maus, Northeast Minnesota Historical Center, Duluth, Minnesota, to the Author, May 17, 2005.

5. Ibid., pp. 128–129.

6. Lamppa, *Minnesota's Iron Country*, p. 167.

7. "The Color of Mesabi Bones," Cited in Thompson, *Iron Range Country*, p. 94.

8. Gus Hall, http://reds.linefeed.org/bios/gushall.htm.

9. Oral History Interview with Gus Hall, Eveleth, Minnesota, October 29, 1980, n.p.

10. Beck and Labadie, *Pride of the Inland Seas*, pp. 130–131.

11. Ibid., p. 127.

12. Ibid., p. 129.

13. Duluth-Superior Harbor Statistics, p. 2.

14. Ibid., p. 2.

15. Beck and Labadie, *Pride of the Inland Seas*, pp. 138–139.

16. Duluth-Superior Harbor Statistics, p. 2.

17. Beck and Labadie, *Pride of the Inland Seas*, p. 106.

18. Ibid., p. 125.

19. Sandvik, *Duluth: An Illustrated History*, p. 104.

20. Ibid., p. 106.

21. Ibid., p. 110.

22. Beck, *Northern Lights: An Illustrated History of Minnesota Power* (Duluth: Minnesota Power, 1985), pp. 81–96.

23. Ibid., pp. 58–61.

24. Sandvik, *Duluth: An Illustrated History*, p. 116.

25. Ibid., p. 102.

26. Beck, *We Remember the Dream: St. Mary's Medical Center, 1888–1988* (Duluth: St. Mary's Medical Center, 1988), p. 18.

27. "U.S. Steel Corporation Will Build Monster Plant in Duluth," *Duluth News-Tribune,* April 2, 1907.

28. Arnold R. Alanen, "Morgan Park: U.S. Steel and a Planned Company Town," in Rick Lydecker and Lawrence J. Sommer, *Duluth: Sketches of the Past: A Bicentennial Collection* (Duluth: American Revolution Bicentennial Commission, 1976), pp. 111–114.

29. Editorial, *Duluth News-Tribune,* January 16, 1910.

Chapter 4

1. Clifton Daniel, John W. Kirshon, and Ralph Berens, eds., *Chronicle of America* (Liberty, Mo.: JL International Publishing) p. 218.

2. Ibid., p. 502

3. Van Brunt, *Duluth and St. Louis County, Minnesota*, v. 1, p. 501.

4. Ibid., p. 501.

5. Ibid., p. 503.

6. *Lake of the Woods County: A History of People, Places and Events* (Baudette, Minnesota: Lake of the Woods County Historical Society, 1997), pp. 257–271.

7. Beck, "The State of the Industry," *Timber Bulletin* 42 (50th Anniversary Edition, 1987): 14.

8. Ibid., p. 13.

9. Hiram Drache, *Taming the Wilderness: The Northern Border Country, 1910–1939* (Danville, Illinois: Interstate Publishers, 1992), p. 28.

10. Ibid., p. 28.

11. Duluth-Superior Harbor Statistics, p. 3.

12. Ibid., p. 3.

13. E-mail, Pat Maus, Northeast Minnesota Historical Center,Duluth, Minnesota, to the Author, June 1, 2005, p. 1.

14. Ibid.

15. Ibid., p. 2.

16. Ibid.

17. Lamppa, *Minnesota's Iron Country*, p. 212.

18. Melvin Dubofsky, *We Shall Be All: A History of the Industrial Workers of the World* (Chicago: Quadrangle, 1969), pp. 322–326.

19. Beck, "Law and Order during the 1913 Michigan Copper Strike," *Michigan History* (Winter 1970): 270–288.

20. Lamppa, *Minnesota's Iron Country*, p. 215.

21. Robert Eleff, "The 1916 Minnesota Miners' Strike Against U.S. Steel," *Minnesota History* (Summer 1988): 74.

22. Sandvik, *Duluth*, pp. 56–57.

23. Arnold L. Alanen, "Morgan Park: U.S. Steel and a Planned Company Town," *Duluth: Sketches of the Past*, pp. 111–126.

24. Beck and Labadie, *Pride of the Inland Seas*, pp. 120–123.

25. Ibid., p. 123.

26. "Hundreds Perish in Flames," *Duluth Sunday News-Tribune*, October 13, 1918, p. 1.

27. Beck, *Northern Lights*, pp. 156–157.

28. "Influenza Cases Number 80; City Fights Plague," *Duluth News-Tribune*, October 16, 1918.

29. Beck, *We Remember the Dream*, p. 37.

30. E-mail, Pat Maus, Northeast Minnesota Historical Center, Duluth, Minnesota, to the Author, May 17, 2005.

Chapter 5

1. The Federal Writers' Program, *The WPA Guide to the Minnesota Arrowhead Country* (St. Paul: Minnesota Historical Society Press, 1988), pp. 161–170.

2. Lamppa, *Minnesota's Iron Country*, pp. 202–203.

3. "Twin Cities-Duluth Flight Is First Ever," *Duluth News-Tribune*, February 23, 1919.

4. "King of Swat Appears Here," *Duluth News-Tribune*, November 7, 1926.

5. *The WPA Guide to the Minnesota Arrowhead Country*, p. 27.

6. *Duluth and St. Louis County*, p. 569.

7. *Hibbing Daily News*, July 4, 1920.

8. "Camera Tours Iron Range With President Coolidge," *Duluth News-Tribune*, August 3, 1928.

9. *The WPA Guide to the Minnesota Arrowhead Country*, p. 209.

10. Beck and Labadie, *Pride of the Inland Seas*, p. 149.

11. Beck, *Northern Lights*, p. 235.

12. "Still Strong Steel," *Time*, July 29, 1929.

13. "Bridge Rises 25 Feet on First Attempt," *Duluth News-Tribune*, March 21, 1930.

14. Maury Klein, *Rainbow's End: The Crash of 1929* (Oxford: Oxford University Press, 2001) p. 221.

15. Ibid., pp. 222–225.

16. Charles A. Beard and Mary R. Beard, *America in Midpassage*, v. III of *The Rise of American Civilization* (New York: The Macmillan Company, 1945), p. 59.

17. Klein, *Rainbow's End*, p. 229. The market would not register another 16-million-share day until April 1968.

18. Beck and Labadie, *Pride of the Inland Seas*, p. 150.

19. Beck, *Northern Lights*, pp. 263–266.

20. Beck and Labadie, *Pride of the Inland Seas*, p. 151.

21. Ibid.
22. "Temperatures Shatter Records," *Duluth News-Tribune,* February 7, 1936.
23. "Cooler is Predicted As Heat Hits 106," *Duluth News-Tribune,* July 14, 1936.
24. "Heat Wave Kills 4,400 in 14 Days," *Duluth News-Tribune,* July 17, 1936.
25. "Tofte Forest Region Blaze Hits Big Area," *Duluth News-Tribune,* July 14, 1936.
26. Beck and Labadie, *Pride of the Inland Seas,* p. 155.
27. Curtis S. Miller, "Organized Labor: A Look Back," in *Duluth: Sketches of the Past,* p. 215.
28. Lamppa, *Minnesota's Iron Country,* p. 221.
29. Ibid., p. 221.
30. Ibid., pp. 221–222.
31. Hall Interview, October 29, 1980.
32. Beck, "A New Day Is At Birth: The 1937 Lumberjack Strike in Minnesota and the U.P.," *Timber Bulletin* (1987): pp. 6–8.
33. Beck, "The Two Elmers," op. cit., pp. 9–12.
34. Ibid., p. 12.
35. Interview with Martin Kuusisto, Duluth, Minnesota, April 25, 1987.
36. Ibid.
37. Beck and Labadie, *Pride of the Inland Seas,* p. 163.
38. Ibid., p. 164.
39. Ibid., p. 165.
40. Ibid., pp. 172–173.

Interlude: 1956

1. *Life in Hibbing,* www.hibbing.org/dylan1/story.html.
2. Albert Woolson, "Last Union Army Veteran Dies," *The New York Times,* August 2, 1956.
3. Ibid.
4. Ibid.
5. "Nation Pays Final Tribute to Last of the Boys in Blue," *Duluth News-Tribune,* August 7, 1956.
6. Albert Woolson, "Last Union Army Veteran Dies," *The New York Times,* August 2, 1956.
7. Gordon Slovut, "Taps for Woolson Echo Across U.S.," *Duluth News-Tribune,* August 7, 1956.
8. Beck and Labadie, *Pride of the Inland Seas,* p. 183.
9. Ibid., p. 183.
10. Ibid., p. 188.
11. Jim Myhers, "Twin Ports Linked to Atlantic Ocean," *Duluth News-Tribune,* May 4, 1959.

Chapter 6

1. The Iron Ore Dilemma, *Fortune,* November 1945, pp. 128–139.
2. Beck, *Northern Lights,* p. 386.
3. Beck and Labadie, *Pride of the Inland Seas,* p. 175.
4. Beck, "The Wilderness Years," *Timber Bulletin* 42 (50th Anniversary Edition, 1987): 56.
5. Ibid., p. 56.
6. Ibid., pp. 56–57.

7. Ibid., p. 57.

8. John Anton Blatnik, *Biographical Directory of the United States Congress*, http://bioguide.congress.gov/scripts/biodisplay.pl?index=B000550.

9. Richard Pearson, "Ex-Rep John Blatnik Dies; Liberal Leader," *The Washington Post,* December 18, 1991.

10. Stephanie Hemphill, "John Blatnik's Contradictory Role," Minnesota Public Radio, October 29, 2003.

11. Ibid.

12. U.S. Senator David Durenberger, "Tribute to John A. Blatnik," *The Congressional Record,* January 29, 1992, p. S557.

13. Beck and Labadie, *Pride of the Inland Seas*, pp. 198–199.

14. Ibid., p. 192.

15. Ibid., p. 193.

16. Ibid., p. 194.

17. Ibid., p. 195.

18. Andrew H. Brown, "New St. Lawrence Seaway Opens the Great Lakes to the World," *The National Geographic Magazine* CXV, no. 3 (March 1959): 299.

19. Ibid., p. 299.

20. Beck and Labadie, *Pride of the Inland Seas*, p. 198.

21. Ralph S. Knowlton, Administrative Officer, Lake Superior Area Office, St. Paul District, U.S. Army Corps of Engineers, Presentation to the Ironwood, Michigan, Kiwanis Club, May 9, 1961, p. 5.

22. Memo, Author to Raymond G. Erickson, Vice President, Minnesota Power, February 27, 1982.

23. Sandvik, *Duluth: An Illustrated History of the Zenith City.*

24. Ibid., p. 75.

25. Ibid., p. 116.

26. Beck, *We Remember the Dream*, pp. 64–66.

27. Ibid., pp. 65–66.

28. Peter Kakela, "Iron Will" (Unpublished manuscript in possession of Author), 2003, p. 39.

29. Ibid., p. 13.

30. Ibid., p. 57.

31. Ibid., pp. 57–58.

32. Ibid., p. 61.

33. Dave Gardner, "Looking Back: A Career Retrospective," *Skillings Mining Review* (August 18, 2001): 4.

34. Ibid., p. 69.

35. Gardner, *Looking Back: A Career Retrospective*, p. 5.

36. Kakela, "Iron Will," p. 70.

37. Ibid.

38. Ibid.

39. Gardner, *Looking Back: A Career Retrospective*, p. 5.

40. News Item, *Inland Seas* (Fall 1973): 211.

41. Beck, "Ox Yokes to Feller Bunchers," *Timber Bulletin* 42 (50th Anniversary Issue, 1987): 35–37.

42. Beck and Labadie, *Pride of the Inland Seas*, p. 210.

43. Ibid., p. 217.

44. Ibid., p. 217.

45. Ibid., p. 220.
46. Ibid., p. 229.
47. Ibid., p. 231.
48. Ibid., p. 238.
49. Jeffrey L. Rodengen, *The Legend of Nucor Corporation* (Ft. Lauderdale, Fla.: Write Stuff Enterprises, 1997), pp. 108–109.
50. Ibid., p. 93.
51. Ibid., p. 93.
52. Beck and Labadie, *Pride of the Inland Seas*, p. 238.
53. Dave Gardner, "Looking Back: A Career Retrospective," *Skillings Mining Review* (August 25, 2001): 4.
54. Ibid.
55. Beck and Labadie, *Pride of the Inland Seas*, p. 240.
56. Ibid., p. 241.
57. Rudy Perpich, www.wikipedia.org/wiki/Rudy_Perpich.
58. "Steel Makes a Comeback," *Duluth News-Tribune & Herald*, September 27, 1987.
59. Gardner, "Looking Back: A Career Retrospective," August 25, 2001, p. 5.

Chapter 7

1. Ed Nelson, St. Louis County Historical Society, Eightieth Anniversary Speech, Duluth, Minnesota, October 2002.
2. Ibid.
3. William E. Culkin to Solon Buck, Persons Invited to Attend Organization Meeting of a St. Louis County Historical Society, November 3, 1922.
4. St. Louis County Historical Society, Audit Report, 1990, n.p.
5. St. Louis County Historical Society *News Quarterly*, January 1956, v. 4, no. 2.
6. Nelson Speech.
7. St. Louis County Historical Society *Newsletter*, June 1983, v. 26, no. 2.
8. St. Louis County Historical Society *Newsletter*, September 1983, v. 26, no. 3.
9. St. Louis County Historical Society *Newsletter*, Spring 1986, v. 29, no. 1.
10. St. Louis County Historical Society *Newsletter*, Summer 1989, v. 32, no. 2.
11. St. Louis County Historical Society *Newsletter*, Winter 1991, v. 34, no. 1.
12. St. Louis County Historical Society *Newsletter*, February 1994, v. 36, no. 2.
13. St. Louis County Historical Society *Newsletter*, June 2004, v. 46, no. 3.
14. www.vets-hall.org

Index

Bold numbers indicate photos.

D

E

F

About the Author

Writer and historian Bill Beck has nearly two decades of experience writing about business and institutional history. He wrote his first history for Minnesota Power in 1985, and has sixty published books to his credit. Recent books include *Pride of the Inland Seas: An Illustrated History of the Port of Duluth-Superior*, published by Afton Historical Society Press in 2004, and *Play On! Celebrating 100 Years of High School Sports in Indiana*, published in 2003 by the Indiana High School Athletic Association. He is currently completing an illustrated centennial history for Minnesota Power. Beck also serves as a field editor for *Indiana Business Magazine, Seaway Review*, and *The Iron Age Scrap Price Bulletin.*

Beck is a 1971 graduate of Marian College and did graduate work in American History at the University of North Dakota. Beck started Lakeside Writers' Group eighteen years ago following ten years as a reporter for newspapers in Minnesota and North Carolina and seven years as the senior writer in the public affairs department at Minnesota Power in Duluth. He currently lives in his hometown of Indianapolis, Indiana.